The *Vinca* Alkaloids

The *Vinca* Alkaloids

Botany, Chemistry, and Pharmacology

Edited by

William I. Taylor

International Flavors and Fragrances
Union Beach, New Jersey

and

Norman Farnsworth

Department of Pharmacognosy and Pharmacology
College of Illinois
Chicago, Illinois

MARCEL DEKKER, INC. New York   1973

MARCEL DEKKER, INC.

95 Madison Avenue, New York, New York 10016

LIBRARY OF CONGRESS CATALOG CARD NUMBER: 73-83859

ISBN: 0-8247-6129-4

Current printing (last digit):

10 9 8 7 6 5 4 3 2 1

PRINTED IN THE UNITED STATES OF AMERICA

CONTENTS

Chapter 2. THE PHYTOCHEMISTRY OF VINCA
SPECIES. Norman R. Farnsworth

CONTRIBUTORS

N. R. Farnsworth, Department of Pharmacognosy and Pharmacology, University of Illinois, Chicago, Illinois

M. Hava, Department of Pharmacology, University of Kansas Medical Center, Kansas City, Kansas

I. Kompiš, Institute of Chemistry, Slovak Academy of Sciences, Bratislava, Czechoslovakia

G. Mark, Gedeon Richter, Ltd. Budapest, Hungary

W. T. Stearn, Department of Botany, British Museum (Natural History), London S. W. 7, England

K. Szász, Gedeon Richter, Ltd. Budapest, Hungary

W. I. Taylor, International Flavors & Fragrance (IFF - R&D), Union Beach, New Jersey

FOREWORD

The medical use of pervenche (French for <u>Vinca</u> <u>minor</u>) has a long history, dating from the 13th century, when it was recommended against nose bleeding and other ailments. Mme. de Sevigne's letter to her daughter (1684) commenting on the merits of a <u>Vinca</u> treatment for an inflamed chest is well known (Lloydia <u>27</u>, 227, 1964). J. J. Rousseau's interest in <u>Vinca</u> (Confessions, Livre 6, p. 226, ed. Pleiade, 1959) is purely botanical and has nothing to do with the healing effects attributed to this plant, for which some of the references are given in Chapter 2 of this book.

The chemical investigation of the different <u>Vinca</u> species developed slowly. Except for a paper in 1934 by Orekhov, Gurevitch and Norkina, the thorough investigation of <u>Vinca</u> alkaloids did not get underway until the early fifties. Since then a wealth of chemical and pharmacological data have accumulated, particularly on the alkaloids of these species.

In June of 1964 a symposium was held in Pittsburgh, as part of the Fifth Annual Meeting of the American Society of

Pharmacognosy. Its organizers have to be highly commended for their endeavors, since this symposium provided a great stimulus for further work. Parenthetically, it was also one of the rare opportunities in recent years when representatives from eastern Europe were able to report upon their important work on Vinca before an essentially American audience. The December 1964 issue of Lloydia carries all of the papers presented at Pittsburgh, and thus the status of Vinca-Catharanthus chemistry up to that date has been clearly defined.

In 1964 it was still a matter of discussion whether Vinca and Catharanthus were two different genera, or one and the same genus. This dispute has now been decided in favor of two genera, and therefore this monograph deals only with the contents of the different Vinca species. It is hoped that a second volume will take care of Catharanthus and its chemical constituents and biological activities.

Seven years have passed since the Pittsburgh meeting. During this time, scientists from more than a dozen countries have contributed toward an enlargement of our knowledge concerning Vinca species, and today some 80 odd alkaloids have been isolated from the different Vinca species. Many important problems in Vinca alkaloid chemistry have been solved, but many remain. Thus, it seems highly appropriate to

publish this critical review, which assembles in a single place

all aspects of the botany, phytochemistry, chemistry and

biological activities of <u>Vinca</u> species, which have been

carried out in so many different laboratories over the past

two decades.

E. Schlittler

PREFACE

The varied character of natural products, and indeed their very existence, pose fundamental questions to scientists. While many books have been published concerning the chemical aspects of natural products, it is an exceptional case when most aspects of a class of substances can be found within the covers of a single volume. We have attempted, with some difficulty, to do this for the <u>Vinca</u> alkaloids, and have brought together the work of experts in the various specialized fields of endeavor. If this volume meets with general acceptance, we would feel encouraged to continue these broad summaries.

This book was written with two groups of readers in mind. The first is typified by graduate or advanced graduates, who are well grounded in organic chemistry, botany or pharmacology. The second group is composed of established researchers. Both groups can use this book either for reference or as a text, especially for the convenience of finding the subject covered so broadly.

This book could not have been reproduced without the dedicated typing efforts of Mrs. Anne-Grete Kreiborg, Mrs. Clarice D. Taylor and Miss Linda Schillace, to whom we are very grateful.

We would also like to acknowledge the assistance of Mrs. W. M. Messmer and Mr. G. H. Aynilian for constructing the formulae for this book.

Jozef Mokrý

(1924 - 1966)

Jozef Mokrý was born on October 19, 1924 in Senec, Czechoslovakia, and was educated in Bratislava. After World War II he matriculated at the Chemical Faculty of the Slovak Technical University, and completed his education four years later with high honors.

At about this time the foundations of the pharmaceutical industry and research were being laid down in Slovakia, and Dr. Mokrý was among those who contributed to its development. At that time he was concerned mainly with the chemistry of xanthine derivatives, an area in which he combined his outstanding technical skill, with theoretical knowledge and organizing ability.

After joining the Slovak Academy of Sciences, he became involved in academic research, and since 1960 his deepest scientific love was the indole alkaloids, and in particular, those derived from <u>Vinca minor</u>. In a brief span of years from 1960-1964, and working under conditions which were far from ideal, he made many significant contributions to this rapidly developing field.

In private life, as in the laboratory, he was always an example of a scientist in every sense of the word. Chemistry was for him more than a profession, but a subject of deep personal interest, to which he dedicated his life. Unfortunately, illness interrupted his efforts and did not allow him to see the fruits of his work.

In dedicating this volume to Dr. Jozef Mokrý, we are paying tribute to both his work and his personality.

The *Vinca* Alkaloids

INTRODUCTION

Ivan Kompiš *

Institute of Chemistry
Slovak Academy of Sciences
Bratislava, Czechoslovakia

Although the genus Vinca has only a small number of

species, it has become the object of intensive research

activities. The concentrated effort of several groups of

chemists, botanists and pharmacologists has afforded com-

prehensive experimental material which has allowed us to add

the genus Vinca to the many other well-explored genera of the

Apocynaceae. The knowledge and conclusions obtained by

examining the constituents of plants from this genus have

contributed greatly to the development of indole alkaloid

chemistry, and have shed light on the complex relationships

of different structural types of indole alkaloids.

Vinca species have a very old tradition in the folk medi-

cine of Europe and of the Middle East. It is no wonder that

* Present address: F. Hoffmann-La Roche and Co., Basel
Switzerland.

1

the first information on these plants found in the literature was

that reporting on the applications of their healing properties.

The aim of utilizing the active principles of plants as drugs

was, and has remained, a powerful stimulus for research in

this area.   The actions and usefulness of Vinca minor (des-

cribed under the name Polemion) were reported by Dioscorides

as early as 50 B.C., and subsequently by Plinius and Galenus.[1]

Since then, the same plant has often been referred to, e.g. in

the works of P.A. Mattioli (1563), Hieronymus Bach (1677), and

in more recent publications.[2,3]   Vinca minor has been an

official drug in the Pharmacopoeia Gallica[4].   In folk medicine

literature there are mentioned several applications for this

plant, such as diuretic, antidysenteric, hemorrhagic and

wound healing.   In addition to these uses, in the middle ages

Vinca minor was claimed to be a potent agent used against

witches and thunderbolt.[2]   Intensive pharmacological research

in recent years has been primarily directed toward the hypo-

tensive action of the alkaloids of this plant.   These actions have

been reviewed,[5] and a detailed treatment of the pharmacology

of the constituents of Vinca species is to be found in chapter 5

of this work.

As will be discussed in greater detail in chapter 1,

according to Pichon,[6] the genus Vinca is classified into three species with five varieties, i.e. Vinca minor L., Vinca major

L., with varieties difformis (Pourr.) Pich. and major (L.)

Pich., and Vinca herbacea Waldst. et Kit., with var. herbacea

Pich., var. libanotica (Zucc.) Pich. and var. sessilifolia

(A.DC.) Pich. Thus far, the following species have been

chemically analyzed, Vinca minor L., Vinca pubescens Urv.

(according to the above nomenclature, this species should be

Vinca major L. var. major Pich.), Vinca difformis Pourr.

(should be Vinca major L. var. difformis (Pourr.) Pich.),

Vinca major L., Vinca herbacea Waldst. et Kit. and Vinca

erecta Rgl. et Schmalh. (should be Vinca herbacea Waldst.

et Kit. var. libanotica (Zucc.) Pich.). To avoid confusion,

we have adhered to the original names used by different

authors for the species studied, even if these were not con-

sistent with the nomenclature of Pichon.[6]

From the chemist's point of view, it seems now to be a

propitious time to summarize the achievements made in this

area. The different species of the genus Vinca have been

examined using many methods of isolation and structure

determination. The rapid development and broad application

of modern physico-chemical methods has substantially

influence the quantity, and most probably the quality of inform-

ation on the alkaloids of this genus. Due to these methods, it can be safely claimed that the major low molecular weight alkaloids of the plants in question are now known. On the other hand, a glance at the current literature points out that research in this area has not terminated, and the isolation of minor low molecular weight alkaloids, their structure determination, and investigation of their pharmacological properties, is continuing.

The very beginning of a chemical approach to the research on Vinca plants dates to 1859, when Lucas isolated an alkaloid fraction from Vinca minor.[7] In 1932, and the following years, there appeared additional papers dealing with the alkaloids and glycosides of Vinca minor,[8] Vinca herbacea[9] and Vinca pubescens.[10] Intensive research in this field started, however, only 20 years later, when interest in the indole alkaloids of the Apocynaceae was stimulated by the isolation and biological activities of reserpine. Thus, Vinca minor was re-examined. Rapid progress can be illustrated by the fact that in 1953 only four alkaloids from this group of plants were known, and these were from Vinca minor and Vinca pubescens. However, by the end of 1968, at least 86 Vinca alkaloids had been isolated, and the structure and stereochemistry for about 68 of these had been deduced.

In addition to alkaloids, which are the most character-istic constituent of Vinca plants, other groups of natural products were also isolated, e.g. glycosides, terpenes,

sterols, phenolic compounds, saponins, hydrocarbons, gums

and pectins. Attempts to isolate cardiac glycosides from

Vinca minor were unsuccessful,[11] but some workers isolated

and identified 3-β-D-glucosyloxy-2-hydroxybenzoic acid.[13]

The triterpene ursolic acid was found in five Vinca species,[14]

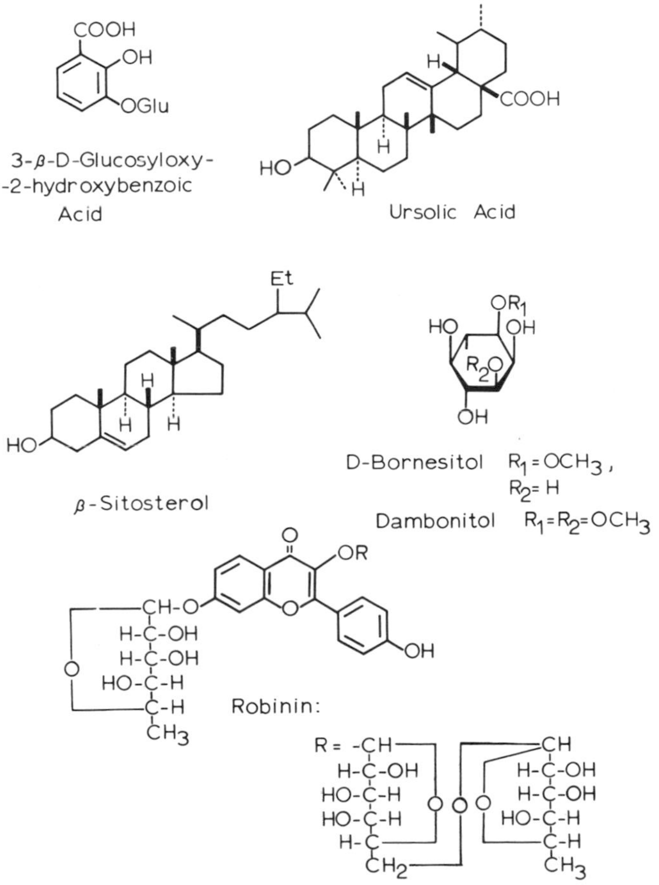

FIG. 1

Structures of non-alkaloidal constituents of Vinca species.

in amounts ranging up to 3.7 per cent. Sterols are represented by β-sitosterol,[15] and of the phenolic compounds, o-pyro-catechuic acid can be mentioned.[16] The esters of myoinositol, namely D-bornesitol and dambonitol,[17] were found in Vinca minor and Vinca major, whereas the flowers from both of these species yielded the flavonoid robinin (kämpferol-7-L-rhamnoside-3-robinoside).[18] The isolation of simple monocarboxylic acids, e.g. formic, propionic, butyric and isobutyric, from some Vinca species has also been recorded[3] (see Fig. 1).

The major effort, however, has been concentrated on the alkaloids of these species. Results of this work have been summarized in several review articles, the most recent one being published in 1965.[19] With regard to the total alkaloid accumulation in Vinca species, it can be said that they are characteristic of most alkaloid-bearing plants grown in temperate zones in producing only moderate amounts, usually within the range of 0.35 to 1.0 per cent, calculated on the weight of air-dried plant material taken. On the other hand, Vinca erecta and Vinca herbacea, grown in subtropical areas, have a much higher alkaloid concentration. For example, the roots of Vinca erecta contain as much as 2.7 per cent of total crude bases.

In a broader context, the alkaloids of <u>Vinca</u> species reveal many common features with the alkaloids from closely related genera of the Apocynaceae (sub-family Plumieroideae), e.g. with the genera <u>Alstonia</u>, <u>Aspidosperma</u>, <u>Amsonia</u>, <u>Rhazya</u>, <u>Geissospermum</u>, <u>Tabernaemontana</u>, <u>Kopsia</u> and <u>Catharanthus</u>.

Until the second half of 1961, knowledge concerning the structure of <u>Vinca</u> alkaloids was based almost exclusively on the results achieved by examining the structures of bases from related genera. (For the structures of the <u>Vinca</u> alkaloids see chapter 2, Figures 1-16). The first <u>Vinca</u> alkaloids whose structures became known were those which had previously been isolated from <u>Rauvolfia</u> and <u>Catharanthus</u> species, viz. reserpinine, sarpagine and akuammine. Up to 1961, structures for only two genuine <u>Vinca</u> alkaloids, viz. vincamajine and vincamedine, had been proposed.[20] In the years 1961 and 1962 the key problem became the structure elucidation of vincamine, the major alkaloid in this genus. First to be clarified, however, was a confusing matter concerned with the synonyms of vincamine and its mixture with 11-methoxy-vincamine (vincine). The many names given to these compounds and mixtures (vincamine, minorine, vincamirine, isovincamine, perivincine) made orientation in this field exceed-

ingly difficult. Then, Bartlett and Taylor's work on eburna-

mine and related alkaloids, [23] and the subsequent structure

elucidation of vincamine, [22] afforded a clue for the structure

determination of a whole series of Vinca alkaloids of the

eburnamine type. Vincamine is obviously the most frequently

observed alkaloid in this genus of plants, as it has been found

not only in Vinca minor, where it is one of the major bases,

but also in Vinca major, Vinca difformis and Vinca erecta

(see Table 1).

In spite of the fact that some alkaloids are more fre-

quently found in the various species of this genus (and also in

other species of the Apocynaceae), the large number of

alkaloids isolated to date enables us to differentiate, to some

extent, between species. Such an approach might be of some

value for a chemotaxonomic study of the Vinca species. How-

ever, one should bear in mind that for some species, e. g.

Vinca minor and Vinca herbacea, only the non-polar fractions

of the total alkaloids have been examined in detail (for a

definition of non-polar and polar fractions, see reference 30).

The polar fraction of Vinca minor represents more than 50

per cent of the total alkaloids. Recently, it was demonstrated

that in addition to quaternary alkaloids in this fraction,

TABLE 1

Alkaloids Occurring in More Than One Vinca Species

Alkaloid	V. minor	V. major var.		V. herbacea var.	
		major	difformis	herbacea	libanotica
Vincamine	+	+	+	-	+
Vincine	+	+	-	-	-
Vincadifformine	+	-	+	-	-
Sarpagine	-	+	+	-	-
Vincamedine	-	+	+	-	-
Vincamajine	-	+	+	-	-
Majdine	-	+	-	+	+
Reserpinine	-	+	-	+	+
Akuammine	-	+	-	-	+
Akuammidine	-	-	+	+	+
Norfluorocurarine	-	-	-	+	+

dimeric bases were also present.[24] These two groups of

alkaloids represent a fertile field for future investigations.

Furthermore, Farnsworth has found, in a paper chromato-

graphic examination of the total alkaloids of Vinca major, at

least 37 alkaloids,[25] but thus far only 10 alkaloids of known

structure and five of unknown structure have been isolated

from this plant. Also, for Vinca erecta, the number of alka-

loids claimed to be present is ca. 30,[26] and only 20 of known

structure, and three of unknown structure, have as yet been

isolated. Thus, from the data presented, it is evident that

any speculations concerning the biogenetic relationships of

alkaloids within the genus Vinca have to made very carefully.

The intensive efforts of the research groups of Battersby,

Arigoni, Leete, Scott and Inouye,[27] directed toward the bio-

synthesis of indole alkaloids, have afforded specific informa-

tion concerning the origin of indole alkaloids. Some of these

biosynthetic experiments were carried out using Vinca major

plants.[27b] Based on a whole series of experiments, trypto-

phan was demonstrated to be one of the building blocks of

indole alkaloids, and the non-tryptophan derived moiety of

these alkaloids was firmly proved by the incorporation of

mevalonic acid, geraniol and sweroside, into vindoline,

reserpinine, catharanthine and 1, 2-(-)-dehydroaspidospermidine, in Catharanthus roseus and in Rhazya stricta (Fig. 2).

Thus, evidence was obtained for the biogenesis of yohimbine,

aspidospermine, and iboga-type bases. The first two of these

types occur in Vinca species.

Most information is available on Vinca minor alkaloids.

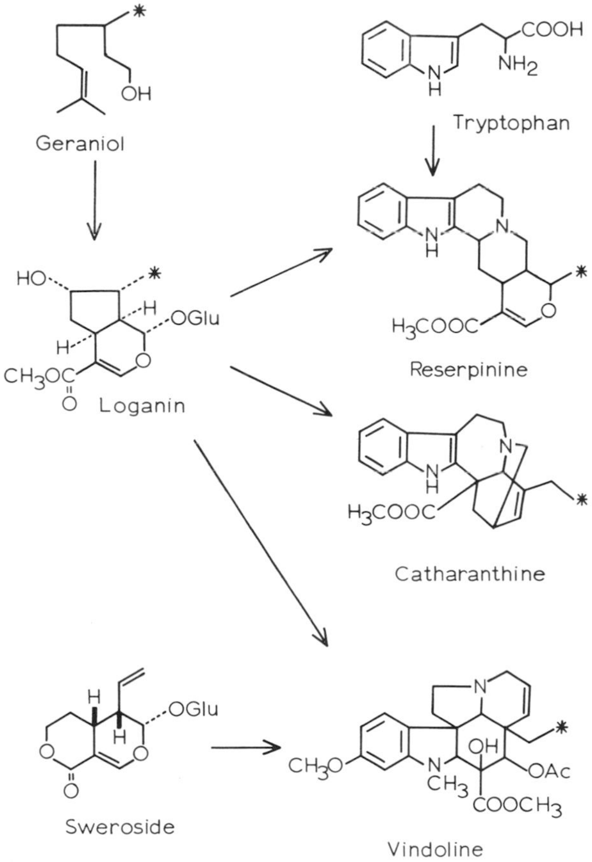

FIG. 2

Biosynthesis of Indole Alkaloids

in Catharanthus and Rhazya Species.

Out of 36 isolated alkaloids, the structures of 30 are known,

with vincamine common to all species except V. herbacea var.

herbacea. With three exceptions, viz. vincamidine[28] (which,

together with vincamine and minovincine represent the major

alkaloids of this plant), vincoridine,[29] and pleiocarpamine

chloride,[30] all of the other alkaloids contain the carbon

skeleton shown in structure II of scheme I. All known alka-

loids with the eburnamine skeleton (with the exception of O-

methyleburnamine[31]), have now been found in Vinca minor.

The most characteristic feature for the bases of Vinca minor

seems to be the presence of tetracyclic indole alkaloids based

on the quebrachamine skeleton [i.e. vincadine, vincaminorine,

vincaminoreine, (+)-N-methylquebrachamine], which to date

have not been found in any other Vinca species.

Of **note is** the stereochemical diversity in the group of

eburnamine, aspidospermine and quebrachamine alkaloids of

Vinca minor. In addition to the occurrence of C-16 epimers,

viz., vincamine-epivincamine, vincaminorine-vincaminoreine

and eburnamine-isoeburnamine, four racemates have also been

isolated, along with their (-)-forms, i.e. (+)-eburnamonine,

N-methyl-(+)-quebrachamine, (+)-vincadifformine, and its

N-methyl derivative, minovine. It has been shown independ-

ently in three different laboratories that the racemization has not taken place during the isolation procedure. Also exhaustive attempts to achieve racemization of the optically active compounds have failed.[30, 33] The occurrence of racemates and/or enantiomers in this class of natural products has been demonstrated in other plants, e.g. the isolation of (+) and (-)-pyrifolidine, as well as (+) and (-)-vincadifformine,[31, 34] and by the isolation of (+)-vincamine and (+)-minovincine from Tabernaemontana ridellii and T. rigida.[34]

Very different are the alkaloid constituents of Vinca major, where the overwhelming majority of bases can be thought of as being derived from "unrearranged" ten-carbon units (see I of Scheme I), which have been isolated in the form of compounds of the ajmaline and sarpagine type.

The alkaloids of Vinca major which contain a "rearranged" non-tryptophan derived part (structure II, scheme I), are represented by vincamine, vincine and (±)-vincadifformine, and these show a distinct relationship with Vinca minor alkaloids. On the other hand, the tetracyclic 2-acylindole alkaloid vincadiffine is a new structural element in the Vinca species. A very close relationship between the constituents of the two varieties of Vinca major, i.e. var. major and var. difformis, is confirmed by the presence of ajmaline and sarpagine-type alkaloids in each of them.

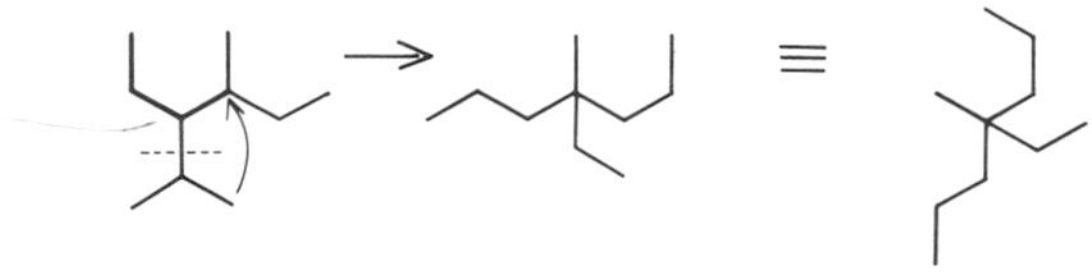

I   II

Scheme I

Terpenoid Skeleta of Vinca Alkaloids

From a historical point of view, Vinca major possesses

a special position because the crystalline bases vinine and

pubescine were the first to be isolated from this plant.[10]

Though these bases have not been re-isolated, some author's

consider them to be reserpinine and majdine (or carapanaubine)

respectively.[32]

More complicated is an evaluation of the alkaloids of Vinca

herbaceae. In this plant, within the group of alkaloids having

an "unrearranged" non-tryptophan derived part of the basic

skeleton (structure I, scheme I), to which belongs the major

alkaloid herbaceine, we also encounter the oxindole alkaloids

majdine, isomajdine and herbaline. Majdine has been isolated

also from Vinca major and from Vinca erecta. Also of

interest is the recent isolation of norfluorocurarine from

Vinca herbacea, which so far is the only alkaloid having the

strychnine skeleton derived from V. herbacea.

Whereas numerous research groups have been involved in the isolation and structural elucidation of the alkaloids thus far discussed, Vinca erecta has to date been a monopoly of the Yunusov-Yuldashev group in Tashkent (UdSSR). This plant contains alkaloids whose structural features are strikingly dissimilar from those of other Vinca species, e.g. norfluorocurarine, and its methoxycarbonyl analogue vinervine. Except for Vinca herbacea and Vinca erecta, the presence of norfluorocurarine has been demonstrated only in Diplorrhynchus condylocarpon.[28] The occurrence of oxindole bases, reserpine and ervine also points out the close relationship of Vinca erecta with Vinca herbacea.

The investigations of Vinca species reviewed in this book documents and emphasizes the aim to discover and exploit biologically active substances from plants which have had an age-long reputation as folk medicines.

The continued use of modern physico-chemical methods is resulting in a rapid increase in our knowledge of different classes of natural products, scattered throughout the current literature. It is hoped that a useful service is being performed in placing all of this information together in a series of monographs on biologically important genera of plants.

REFERENCES

1. G. Madaus, Lehrbuch der biologischen Heilmittel, Vol. 3,
    Leipzig, 1938.

2. L. Kroeber, Das Neuzeitliche Kräuterbuch, Vol. 2,
    Hippokrates Verlag, Stuttgart, 1947.

3. E. Perrot, Matieres premieres usuelles du regne
    vegetal, Vol. 3, Masson et Co., Paris, 1944.

4. Pharmacopeé francaise Editio VIII., 1965.

5. F. H. L. van Os, Pharm. Weekblad 96, 966 (1961).

6. M. Pichon, Bull. Mus. Hist. Nat., 23, 439 (1951).

7. H. Lucas, Arch. Pharm., 147, 147 (1859).

8. F. Rutishauser, Compt. Rend. 195, 75 (1932).

9. a. I. Vintilesco and N. I. Ioanid, Bull. Soc. Chim. Biol.,
       15, 63 (1933).
    b. I. Vintilesco and N. I. Ioanid, Bull. Soc. Chim.
       Romania, 14, 12 (1932).

10. A. Orechov, H. Gurevich and S. Norkina, Arch. Pharm.,
    272, 70 (1934).

11. E. S. Zabolotnaya, Trudy VILLAR Moskva, 10, 29 (1950).

12. E. Schlittler and E. Furlenmeier, Helv. Chim. Acta, 36,
    2017 (1953).

13. F. E. King, J.H. Gilks and M. W. Partridge, J. Chem.
    Soc., 1955, 4206.

14. a. V. minor: J. LeMen and J. Pourrat, Ann. Pharm.
       Franc., 10, 349 (1952).
    b. V. major: J. LeMen and Y. Hammouda, Ann. Pharm.
       Franc., 14, 344 (1956).

15. a. D. Zachystalová, O. Štrouf and J. Trojánek, Chem.
       Ind. (London), 1963, 610.
    b. N. R. Farnsworth, H. H. S. Fong, R. N. Blomster
       and F. J. Draus, J. Pharm. Sci., 51, 217 (1962).

16. R. K. Ibrahim, Naturwissenschaften, <u>50</u>, 734 (1963).

17. V. Plouvier, Compt. Rend. , <u>253</u>, 3047 (1961).

18. J. Rabaté, Bull. Soc. Chim. Biol. , <u>15</u>, 130 (1933).

19. W. I. Taylor in the book: R. H. F. Manske, The
     Alkaloids, Vol. VIII, Academic Press, New York, 1965.

20. J. Gosset, J. LeMen and M. -M. Janot, Bull. Soc. Chim.
     France, <u>1961</u>, 1033.

21. a. Z. Čekan, J. Trojánek and E. S. Zabolotnaya,
        Tetrahedron Letters, <u>1959</u>, 11.
     b. J. Mokrý and I. Kompiš, I. Conference on Cardio-
        vascular Active Substances, Smolenice, 1959.

22. a. J. Trojánek, O. Štrouf, J. Holubek and Z. Čekan,
        Tetrahedron Letters, <u>1961</u>, 702.
     b. J. Mokrý, I. Kompiš, J. Suchý, P. Šefčovič and Z.
        Voticky, Chem. Zvesti, <u>17</u>, 41 (1963).
     c. M. Plat, D. Dohkac Manh, J. LeMen, M. -M. Janot,
        H. Budzikiewicz, J. M. Wilson and C. Djerassi, Bull.
        Soc. Chim. France, <u>1962</u>, 1082.

23. M. F. Bartlett and W. I. Taylor, J. Am. Chem. Soc. ,
     <u>82</u>, 5941 (1960).

24. I. Kompiš and E. Grossmann, to be published.

25. N. R. Farnsworth, Lloydia, <u>24</u>, 105 (1961).

26. N. Abdurakhimova, P. Kh. Yuldashev, S. Yu. Yunusov,
     Khim. Prir. Soedin. , <u>1</u>, 224 (1965).

27. a. A. R. Battersby in the book: Chemistry of Natural
        Products, Vol. 4, Butterworths, London, 1967 and
        the references therein.
     b. H. Goeggel and D. Arigoni, Experentia, <u>21</u>, 369 (1965).
     c. H. Inouye, S. Ueda and Y. Takeda, Tetrahedron Letters
        <u>1968,</u> 3413.

28. References concerning the isolation and structure deter-
     mination of the mentioned <u>Vinca</u> alkaloids are given in
     Chapters 2 and 3.

29. I. Kompiš and J. Mokrý, Collection Czech. Chem. Comm.
     1968, (in press).

30. J. Mokrý and I. Kompiš, Lloydia 27, 428 (1964).

31. M. Hesse, Indolalkaloide in Tabellen, Springer Verlag,
     Berlin, 1964 and Ergänzungswerk, 1968.

32. a. J. L. Kaul and J. Trojanek, Lloydia, 29, 26 (1966).
     b. M.-M. Janot, J. LeMen and Y. Hammouda, Compt.
        Rend., 243, 85 (1956).
     c. B. Pyuskyulev, I. Kompiš, I. Ognyanov and G.
        Spiteller, Collection Czech. Chem. Comm., 32, 1289
        (1967).

33. J. Mokrý, I. Kompiš, P. Šefčovič and S. Bauer,
     Collection Czech. Chem. Comm., 28, 1309 (1963).

34. M. P. Cava, S. S. Tjoa, Q. A. Ahmed and A. I. DaRocha,
     J. Org. Chem., 33, 1055 (1968).

CHAPTER 1

A SYNOPSIS OF THE GENUS VINCA
INCLUDING ITS
TAXONOMIC AND NOMENCLATURAL HISTORY

William T. Stearn

Department of Botany
British Museum (Natural History)
London, England

I. INTRODUCTION

The plants commonly known in English as periwinkles
divide into two very distinct groups which must be treated as
separate genera. One is Vinca L. exemplified by V. major L.
and V. minor L., the other is Catharanthus G. Don (Lochnera
Reichenb.) exemplified by C. roseus (L.) G. Don (Vincarosea
L.). They differ in habit, floral structure and distribution,
as pointed out, for example, by Lawrence[1] in 1959 and Stearn[2]
in 1966, and in biochemistry, as pointed out by Farnsworth[3] in
1961; they also differ cytologically. Vinca comprises the periwinkles of the temperate zone, native to Europe, western and
central Asia and naturalized in North America. Catharanthus

belongs to the Old World tropics; namely Madagascar and India, but one species, the ornamental C. roseus from Madagascar, naturalizes so readily that it now has a pantropical range. The major differences between the two genera become apparent on slitting open and comparing flowers of Vinca minor or V. major and Catharanthus roseus. In Vinca the corolla tube widens gradually; the filaments of the stamens are bent forward, then bent backward, somewhat like a knee; their anthers are each crowned with a short hairy flap-like appendage, and they surround the conical tip of the style head. In Catharanthus the corolla tube is almost evenly cylindric; the filaments of the stamens are so short in C. roseus that the anthers are attached almost directly to the corolla tube although they are longer in C. lanceus; the anthers have no terminal appendages and they come together in a conical fashion a little distance above the style head. These differences are illustrated by Lawrence[1] and Stearn[2]; there also are others; thus Pichon[4] in 1948 listed 34 points of distinction between the two genera.

Regarding Vinca proper, for which he used the old designation Vinca Pervinca, the herbalist John Gerard wrote in 1597 in The Herball or Generall Historie of Plantes:

"There be divers sortes or kindes of Pervinkle,

whereof some be greater, others lesser, some with
white flowers, others purple and double, and some
of a faire blew skie colour".

He knew best the species now known as the Lesser Periwinkle

(Vinca minor L.) which

"hath slender and long branches trailing upon the
ground, taking hold here and there as it runneth,
small like unto rushes, with naked or bare spaces
between joint and joint. The leaves are smoothe,
not unlike to the Bay leafe but lesser. The flowers
growe hard by the leaves, spreading wide open,
composed of five small blew leaves".

Likewise grown in London gardens then were "a kind hereof

bearing white flowers", i.e. V. minor 'Alba', and

"another with purple flowers, doubling itselfe somewhat in the middle, with smaller leaves",

i.e. V. minor 'Multiplex'. Gerard also recorded another

sort greater than the rest,

"which is called of some Clematis Daphnoides, of the
similitude the leaves have with those of the Bay. The
leaves and flowers are like unto the precedent, but
altogether greater".

This second species is now known as the Greater Periwinkle

(Vinca major L.).

These two species had been grown in England long before

Gerard's time, although his description is the first adequate

one in English. Indeed, since both easily become naturalized

as escapes from gardens and may survive when gardens have

been abandoned, showing great persistence when once estab-

lished, they may in some places be relics of the Roman

occupation of Britain which lasted from A. D. 43 to A. D. 407.

Certainly the association of periwinkles with man goes back

over two thousand years. Pliny in the first century A. D. wrote:

> "Nam vincapervinca semper viret, in modum lineae
> foliis geniculatim circumdata"

(Plinius, Hist. 21 Cap. 39), a passage very freely rendered by

Philemon Holland in 1634:

> "As for the Pervincle, it continueth fresh and green
> all the year long, this hearbe windeth and runneth
> too and fro with her fine and slender twigges in the
> manner of threads or laces and those beset by leaves
> two by two in order, at every knot or joint"

(Holland, P. , Hist. 2:92, 1634). The name vincapervinca is

evidently derived from vincio, "bind, wind about", and per,

"through", and alludes to the long slender flexible shoots inter-

twined when making garlands and wreaths, for which, having

evergreen leaves, they supplied material all the year round.

From this Latin designation are derived the Late Latin pervinca,

the Middle English pervinke, the 16th century English pervinkle,

perwyncle, perewinkle, and periwinkle, the Italian pervinca

and provenca, Sardinian pruinca and poinca, Catalan vincla-

pervincla, French pervenche, German Berwinkel and Barwinkel,

as well as the generic name Vinca adopted by Linnaeus.

There also exists a series of vernacular names all connected with death, e.g. the French herbe aux mortes,
Provencal vioouleto dey morts, Italian fior da morto, mortini,
viola da morti, German Totenkraut, Totenveieli. These either
allude to the planting of periwinkle in graveyards or to the use
of periwinkle shoots in wreath-making, for during the Middle
Ages condemned persons on their way to execution sometimes
wore a crown of periwinkle. This was evidently a widespread
custom. None of such names is English, but that this use of
periwinkle once existed in Britain is evident from the ca. 1306
Execution of Sir S. Fraser,

"Y-fettered were ys Legges under his horse wombe...
A gerland of pervenke set on ys heved"

(quoted from Oxford Engl. Dict. 7:707; 1909) and Lydgate's
line of ca. 1430-40,

"Other with pervinke made for the gybet".

Planting of periwinkles in graveyards ensured an ever-ready
supply for this gruesome purpose. Certainly the present wide
distribution in Europe of V. minor and V. major is partly the
result of former cultivation; consequently their natural ranges
are ill-defined. Both are certainly native in southern Europe
but many northern stations are questionable.

John Ray referred to the periwinkles under the heading

"De Clematide Daphnoide seu Vinca pervinca" in his Historia

Plantarum 2:1091 (1788) and described the two species under

Caspar Bauhin's names Clematis Daphnoides minor and

Clematis major but gave no generic definition. This was

supplied by Tournefort in his Elemens de Botanique 99, t. 45

(1694), of which a Latin version appeared in his Institutiones

Rei Herbariae 1:119; t. 45 (1700);

> "Pervinca est plantae genus, flore monopetalo,
> infundibuliformi, quasi hypocratoriformi et multi-
> fido: ex cujus calyce surgit pistillum infimae floris
> parti adinstar clavi infixum, quod deinde abit in
> fructum ex duabus siliquis constantem, semine foetis
> oblongo, plerumque cylindraceo et sulcato".

Tournefort replaced the earlier two-word generic designations

Clematis Daphnoides and Vinca Pervinca by the one-word

Pervinca and distinguished in 1694 simply two species Pervinca

vulgaris latifolia (now Vinca major) and Pervinca vulgaris

angustifolia (now Vinca minor) . In 1700, however, he recog-

nized several variants within each, i. e. a blue-flowered form

(flore caeruleo) and a white-flowered form (flore albo) of his

latifolia and forms with single blue, white and purple flowers

and double blue, purple and variegated flowers of his angusti-

folia and also a form of each with variegated leaves, in all

three variants of V. major and seven of V. minor. Tournefort's

plate 45 represents V. major, the species designated by Britton and Brown in 1913 as lecototype of the genus Vinca.

Linnaeus in his Genera Plantarum 63 no. 180 (1737)
shortened Tournefort's Pervinca to Vinca and, as he adopted
this in Species Plantarum 1:209 (1753) and Genera Plantarum
5th ed. 98 no. (1754), which together constitute the starting
point of modern botanical nomenclature, Vinca is necessarily
the correct generic name. He distinguished two species in
1753, i.e. V. minor (of Germany, England and France) having
caulibus procumbentibus, foliis lanceolato-ovatis and V. major
(of Southern France and Spain) having caulibus erectis, foliis
ovatis, but he nevertheless doubted their distinctness, the
reason for this being simply that he knew well V. minor, which
is a very hardy species, but not so well V. major, which is
more easily damaged in a northern winter and hence not successfully cultivated so far north. More southern cultivators,
notably Philip Miller at the Chelsea Physic Garden, in the
London area, had no such doubts. In the fourth edition (1754)
of his Gardener's Dictionary Abridged Miller continued to use
the Tournefortian name Pervinca but in the seventh edition
(1759) of his Gardener's Dictionary he adopted the Linnaean
name Vinca; to Linnaeus's two species, both European, he
now added a third, introduced from Madagascar, which he

had described and illustrated in his Figures of the Most Beau-

tiful, Useful and Uncommon Plants 2:124, t. 186 (1757) as

Vinca foliis oblongo-ovatis integerrimis etc. and which Lin-

naeus himself named V. rosea in 1759. This third species

later became the type of the generic names Catharanthus,

Lochnera and Ammocallis.

The next addition to the genus was V. difformis, which

the Abbé Pierre André Pourret came across in 1783 at Font-

froide, Aude, on a botanizing journey from Narbonne, southern

France, to the Pyrenees. This species is closely related to

V. major but in northern cultivation is less hardy. Its distri-

bution is West Mediterranean, extending from Italy and southern

France over Spain and Portugal to northwestern Africa. Pourret

published his new species in the Hist. Mem. Acad. Roy. Sci.

Inscript. & Belles Lettres de Toulouse 3:337 (1788) with a

brief diagnosis which sufficed, however, to distinguish it from

the already known V. major and V. minor: "1217. Vinca (dif-

formis) foliis ovatolanceolatis glabris; floribus terminalibus

irregularibus, calyce inaequali tubo longiore. A Frontfroide".

The Toulouse Histoire et Mémoires is a comparatively rare

periodical. Not surprisingly, Link and Hoffmannsegg over-

looked Pourret's name when they described and illustrated

the same species occurring near Lisbon as V. media in their

Flore Portugaise 1:376, t. 70 (1813-20) and Bertoloni overlooked

theirs as well when he named it V. acutiflora in his Flora

Italica 2:751 (1836).

The above three species are evergreen. Publication by

Waldstein and Kitaibel in their Descr. Icones Plant. Rar. Hung.

1:8, t. 9 (1799) of V. herbacea from the neighborhood of Buda-

pest, Hungary, but also recorded from the Bihar Mountains,

Roumania, showed that this habit was not a generic character

since V. herbacea dies back to a perennial rootstock every

year, its long trailing shoots being annual. The same species

was collected by the English traveller and scholar Edward

Daniel Clarke in April 1802, on Mount Haemus, southern

Bulgaria, when journeying from Istanbul to Rustchuk, and he

named it V. pumila in his Travels 2:iii:555 (1816).

Further major extensions of the known generic range were

made by the publication in 1822 of V. pubescens D'Urville (now

considered conspecific with V. major) from Sukhumi, western

Caucasus, in 1840 of V. libanotica Zuccarini (now considered

conspecific with V. herbacea) from Lebanon, and in 1879 of

V. erecta Regel & Schmalhausen from the Fergana valley,

Kirgiz S.S.R., Central Asia.

In revising the family Apocynaceae for de Candolle's

Prodromus, Alphonse de Candolle necessarily surveyed the

genus Vinca, of which he published an account in vol. 8:381-
384 (1844). He divided the genus into three sections, i. e. I.
Lochnera (with Catharanthus a synonym) including 1. V. rosea
L. and 2. V. lancea Bojer ex A. DC.; II. Cupa-Veela including
3. V. pusilla Murray; III. Pervinca including 4. V. pubescens
D'Urville, 5. V. sessifolia A. DC., 6. V. herbacea Waldst.
& Kit., 7. V. libanotica Zucc., 8. V. minor L., 9. V. media
Link & Hoffmg. and 10. V. major L. This remained the only
published survey of Vinca as a whole until 1951 when Marcel
Pichon published his "Les espèces du genre Vinca" in Bull.
Mus. Nat. Hist. Nat. Paris II. 23:439-444 (1951). The "full syno-
nymic enumeration of the periwinkles" which I announced with
teen-age optimism in 1930 (Gard. Chron. 88:516; 20 Dec. 1930)
was never issued although it has been used in preparing the
present account. Pichon separated Catharanthus from Vinca
proper, as many authors had done earlier, and then dealt
drastically with Vinca itself, within which he recognized only
three species, i. e. V. major (including V. difformis and V.
pubescens), V. minor and V. herbacea (including V. libanotica
and V. erecta). He stated candidly that, although

> "il existe des formes bien définies qui s'observent en
> peuplements homogènes",

he had deliberately neglected these in his revision.

De Candolle's broad concept of the genus Vinca was ac-

cepted by Bentham and Hooker in their Genera Plantarum

2:703 (1876), who divided it into two sections, i.e. Pervinca

and Lochnera. K. Schumann in Engler & Prantl, Pflanzen-

familien IV. 2:145 (1895), gave these generic rank as Vinca

and Lochnera. The first, however, to recognize these two

groups as separate genera was Ludwig Reichenbach who

in 1828 proposed the generic name Lochnera, under which he

cited Vinca rosea L. Unfortunately Reichenbach provided

neither a generic description nor a statement of differences

between Lochnera and Vinca. The name Lochnera accordingly

remained a nomen nudum (a name published without accom-

panying description or diagnosis or reference to either and

hence not validly published accordingly to the International

Code of Botanical Nomenclature) until Endlicher provided a

description in August 1838 (Gen. Pl. 583 no. 3406). Meanwhile

however, George Don the younger had validly published the

name Catharanthus some time between 1835 and April 1838.

Don's name thus has priority over Lochnera Rchb., which is

also invalid as a later homonym of Lochneria Scopoli (1777);

for a discussion of these nomenclatural matters, see Stearn

in Lloydia 29:196 (1966). This separation of Vinca and Catha-

ranthus is now generally accepted.

II. CHROMOSOME NUMBERS

The first recorded cytological investigation of <u>Vinca</u> seems to be that of W. W. Finn working at Kiev, Ukraine, on <u>Vinca</u> <u>minor</u> and <u>V</u>. <u>herbacea</u>. He recorded n=23 for <u>V</u>. <u>minor</u> in 1928 (<u>Ber. Deutschen Bot. Ges</u>. 46:246) and stated that <u>V</u>. <u>herbacea</u> agreed with it cytologically, hence had n=23 also. These counts were presumably based on Ukrainian material.

In 1937 Fernanda Pannocchia-Laj, using Italian material at Pisa, confirmed 2n=46 for <u>V</u>. <u>minor</u> and recorded 2n=92 for <u>V</u>. <u>major</u> (<u>Nuovo Giorn. Bot. Ital</u>. N.S. 44:340). A year later, in a very detailed paper (<u>Nuovo Giorn. Bot. Ital</u>. N.S. 45:157-187; 1938), she repeated the counts 2n=46 for <u>V</u>. <u>minor</u> and 2n=92 for <u>V</u>. <u>major</u>, from plants cultivated in the Pisa Botanic Garden. Her record of 2n=46 for <u>V</u>. <u>difformis</u> was based, however, on plants wild in Sardinia, which are taxonomically different from true <u>V</u>. <u>difformis</u> and are here named <u>V</u>. <u>difformis</u> subsp. <u>sardoa</u>.

In 1940 Wray Bowden, using plants cultivated in North America, likewise established 2n=46 for <u>V</u>. <u>minor</u>, <u>V</u>. <u>minor</u> 'Alba', <u>V</u>. <u>minor</u> 'La Grave' and <u>V</u>. <u>herbacea</u>, and 2n=92 for <u>V</u>. <u>major</u> (<u>Am. J. Bot</u>. 27:360).

In 1941 J. P. Rutland confirmed 2n=46 for <u>V</u>. <u>minor</u> and 2n=

92 for V. major (New Phytol. 40:211) on plants occurring in
England.

In 1945 Wray Bowden added 2n=46 for V. difformis (from
Coimbra, Portugal) and V. minor 'Multiplex' and 2n=92 for
V. major 'Variegata'. In 1969 Abilio Fernandes confirmed
2n=46 for plants of V. difformis growing around Coimbra, the
type region of V. acutiflora (Fernandes in litt.)

Further counts of chromosome numbers in Vinca have
been made on my behalf during 1968 and 1969 by Keith Jones
at the Jodrell Laboratory, Kew. These gave 2n=46 for 5
variants of V. minor, including 'Alba', 'Argenteovariegata',
'Atropurpurea', 'La Grave' and 'Plena'; for V. herbacea
introduced by J. D. A. Stainton from Asia Minor and V. herbacea
introduced by W. K. Asler above the Cedars, Kadisheh Valley,
above Becharre, Lebanon, the type-region of V. libanotica

2n=about 46 for V. difformis, both the old cultivated stock
of unknown origin and a new introduction by J. do Amaral
Franco from Lisbon, Portugal; 2n=about 92 for V. major
'Oxyloba', the old cultivated stock described by me as
V. major var. oxyloba in 1930; and 2n=92 for V. major subsp.
hirsuta, newly introduced as V. pubescens from the Main
Botanic Garden, Moscow, U.S.S. R. and from the Sukhumi

Botanic Garden, Georgia, U.S.S.R., the type locality of V.

pubescens. The number 2n=16, determined already by several

workers, for Catharanthus roseus (Lochnera rosea), was

confirmed on three forms of the species.

The results of these 35 counts establish 2n=46 for V. dif-

formis, V. minor and V. herbacea and 2n=92 for V. major,

the chromosome numbers of V. balcanica and V. erecta have

yet to be determined. Although V. major (2n=92) is closely

allied to V. difformis (2n=46) it certainly cannot be considered

a direct polyploid of either.

III. GEOGRAPHICAL DISTRIBUTION

The genus Vinca, like Apocynum itself, is a north tempe-

rate outlier of the predominantly tropical and woody family

Apocynaceae and has its nearest ally in the genus Catharan-

thus of Madagascar and India. The main area, within which

species of Vinca are native, extends eastwards from Morocco,

Algeria, Portugal, Spain and France over central and southern

Europe to southwestern European Russia, including the Crimea

and the north Caucasus, and across Asia Minor southward to

Palestine, Syria and Iraq, and eastward to the Caucasus and

northern Iran. Inside this area the ranges of the species differ

but overlap, despite different ecological preferences and tolerances. Thus the range of <u>Vinca</u> <u>minor</u> overlaps in southwestern

Europe that of <u>V</u>. <u>difformis</u> and <u>V</u>. <u>major</u> (subsp. <u>major</u>), and

in central and eastern Europe, that of <u>V</u>. <u>herbacea</u>. Each,

however, occurs also in an area from which the others are

absent, e. g. <u>V</u>. <u>difformis</u> in northwestern Africa and most of

the Iberian Peninsula, <u>V</u>. <u>major</u> in southern Italy, <u>V</u>. <u>minor</u>

in much of western and central Europe, and <u>V</u>. <u>herbacea</u> in

western Asia, where it reaches its easternmost station in

northern Iran. Outside this more or less geographically con-

tinuous area there is only one species, <u>V</u>. <u>erecta,</u> isolated far

to the east of it in the mountains of central Asia (Turkistan and

Afghanistan).

These different ranges correlate with different habits of

growth. <u>V</u>. <u>difformis</u> and <u>V</u>. <u>major</u> are comparatively large-

leaved evergreen species, particularly liable to damage by

hard winters, and they inhabit areas of mild usually frost-free

climate, where they grow in shady places. <u>V</u>. <u>minor</u>, a small-

leaved dwarfer evergreen species, prefers montane regions,

ascends to 4300 ft (1320 m.) on the Alps and extends much

further north in Europe. It is, however, essentially a plant

of light woodland. <u>V</u>. <u>herbacea</u>, which dies down completely

to ground level, prefers open sunny places and is able to sur-

vive under conditions of drought or cold which would kill other

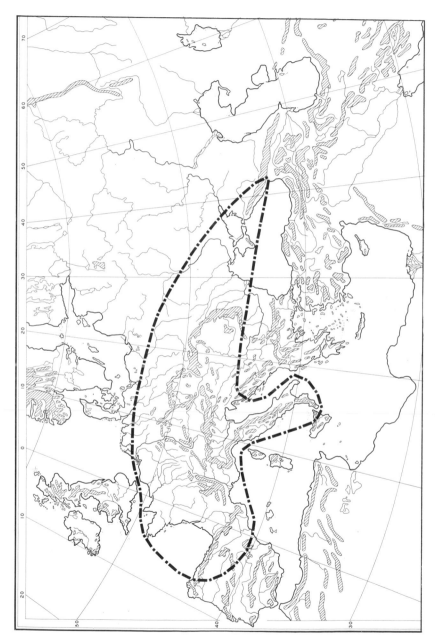

FIG. 1.   Distribution of V. minor L. (Base map: Goode's Series, Univ. of Chicago, No. 124).

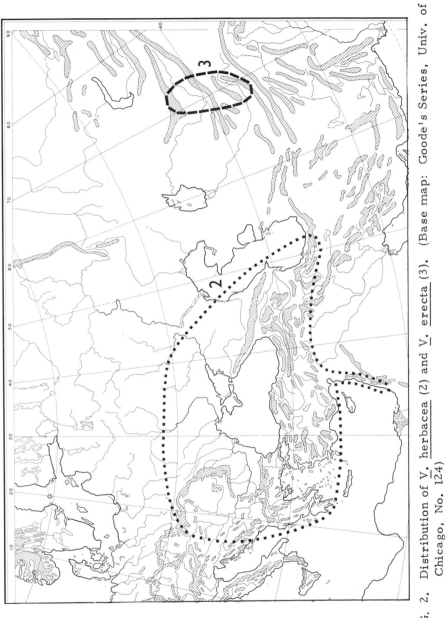

FIG. 2.   Distribution of <u>V</u>. <u>herbacea</u> (2) and <u>V</u>. <u>erecta</u> (3).   (Base map: Goode's Series, Univ. of Chicago, No. 124)

species. V. erecta, closely allied to this, is likewise completely herbaceous. It ascends to 6000 ft (1850 m.) in Afghanistan.

To Pannochia-Laj belongs the credit for initiating specu-
lation about the phylogeny of the species of Vinca on cytological evidence. From other members of the genus, V. major
diverges in its high chromosome number (2n=92) and is evidently tetraploid, since the other species, V. minor, V.

herbacea, V. difformis and ssp. sardoa ("V. difformis" sensu
Pannochia-Laj) have half that number (i. e. 2n=46). These
however, divide into two groups, V. minor and V. herbacea
with very small chromosomes and V. sardoa and V. difformis
with stouter chromosomes. V. minor, being diploid and evergreen with woody shoots, would seem nearest to the primitive
ancestor of the genus. From such a species, V. herbacea
could have evolved without change in number and size of
chromosomes, its completely herbaceous habit enabling it to
colonize the wide area in southeastern Europe and western
Asia with steppe and arrid open scrubland. It has certainly
become the most variable species. The nomenclatural type,
described from Hungary in the northwestern part of its range,
has comparatively small leaves which are broadest about the
middle, and are scabrid-margined. In the southwestern part
of its range, i. e. Lebanon, Palestine, etc., the leaves tend

to be broadest below the middle. The Lebanon population was

named V. libanotica. More extreme variants with much broader,

narrowly ovate or ovate leaves, occur in eastern Asia Minor

and Iraq. To these, the names V. sessilifolia, V. bottae and

V. haussknechtii have been applied. There is some associated

variation in the calyx, which tends to be about 4-5 mm long in

the north and up to 7-8 mm long in the southeast. Thus this

species has proceeded towards regional differentiation without,

however, the regional populations (apart from the well-isolated

central Asiatic V. erecta) achieving distinction through the

elimination of intermediate forms.

V. major presumably evolved direct from a minor-like

ancestor with a doubling of chromosomes. V. balcanica, now

restricted to a small area in the Balkan Peninsula, possibly

represents such an ancestral stage. Unfortunately, the num-

ber and size of its chromosomes are unknown. Then V. major

during a period of mild climatic conditions, presumably inter-

glacial, spread westward and eastward but, being unable to

endure severe cold, became broken during a glacial period

into two widely separated populations. One of these (subsp.

hirsuta) being restricted to the warm Pontic lowland of the

Caucasus and adjacent Asia Minor, and the other (subsp.major),

which retained or evolved greater variability, spreading across

southern France and much of Italy.

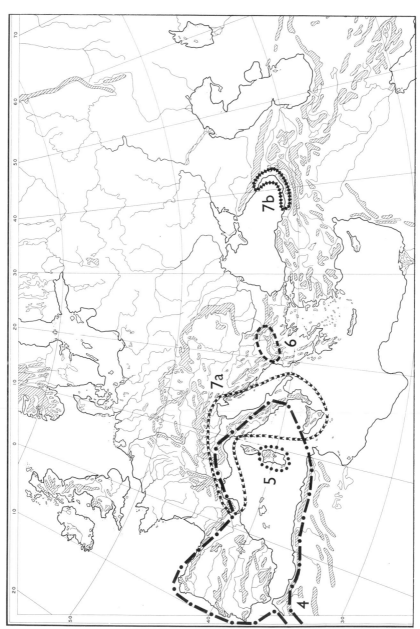

FIG. 3. Distribution of V. difformis ssp. difformis (4), V. difformis ssp. sardoa (5), V. balcanica (6), V. major ssp. major (7a), and V. major ssp. hirsuta (7b). (Base map: Goode's Series, Univ. of Chicago, No. 124)

Since <u>V</u>. <u>major</u> with 2n=92 comes so close in its large
leaves and flowers and habit of growth to <u>V</u>. <u>difformis</u> and
ssp. <u>sardoa</u>, each with 2n=46, it might be assumed to be a tetraploid derivative of these, but for their chromosomes being
larger. All three of these species are much larger than <u>V</u>.
<u>minor</u> and thus show gigantism in relation to this, a state, as
emphasized by F. Schwanitz[5] in 1957, which can result not only
from an increase in the number of chromosomes, i.e. poly-
ploidy, but also from an increase in the size of the chromosomes
without increase in number, i.e. cryptopolyploidy. The latter
occurs, for example, in various cultivated plants. If, then,
<u>V</u>. <u>difformis</u> and ssp. <u>sardoa</u> are cryptopolyploids, as indicated
by Pannochia-Laj, neither can be directly ancestral to <u>V</u>.
<u>major</u>, which has chromosomes more like those of <u>V</u>. <u>minor</u>,
but both represent parallel evolution from the same stock.
Presumably ssp. <u>sardoa</u> achieved its few distinctive characters
through isolation on Sardinia. <u>V</u>. <u>difformis</u>, less cold-tolerant
than <u>V</u>. <u>major</u>, has certainly spread over a wider and more
western area, presumably passing from the Iberian Peninsula
into northwestern Africa.

The genus merits detailed cytotaxonomic investigation
based on living plants of wild origin from many localities.

IV. VINCA L.

Vinca L., Sp. Pl. 1:209 (1753), Gen. Pl. 5th ed. 98 (1754)

Endlicher, Gen. Pl. 583 (1838); K. Schumann in Engler & Prantl,

Pflanzenfam. IV. 2:145 (1895); Woodson in N. Amer. Fl. 29:125

(1938); Pichon in Mém. Mus. Nat. Hist. Nat. Paris N. S. 27:206 (1948).

Pervinca Miller, Gard. Dict. Abridg. 4th ed., 3:art. Per-

vinca (1754); Adanson, Fam. Pl. 2:172 (1763); Caruel in Parlatore,

Fl. Ital. 6:708 (1886).

Vinca sect. Pervinca A. D. C. in D. C., Prodr. 8:382

(1844).

Low evergreen subshrubs or herbaceous perennials.

Leaves opposite, entire. Flowers axillary, solitary, ebrac-

teate, pedunculate. Calyx segments (sepals) 5, shortly fused

at base, narrow, almost equal, without internal squamellae

or glands. Corolla infundibuliform; tube gradually expanded,

with a zone of hairs above the insertion of the stamens in the

upper part; segments (lobes) 5, horizontally spreading, oblique,

in bud overlapping to the left (seen from the outside), joined

at base by a low ridge. Stamens 5, inserted about the middle

of the corolla tube, included; filaments longer than the anthers,

flattened, abruptly bent at the base; anthers basifixed, not

connivent; connective broad, expanded above the thecae into

a hairy flap-like appendage; pollen smooth, with 3 grooves.

Disc consisting of 2 ovoid fleshy nectar-secreting scales (disc

FIG. 4.

FIG. 4.
Diversity in form
of corolla in <u>Vinca</u>.

A. <u>V</u>. <u>minor</u> (type)

B. <u>V</u>. <u>minor</u> 'Plena'

C. <u>V</u>. <u>herbacea</u>
   (from Hungary)

D. <u>V</u>. <u>major</u> ssp.
   <u>major</u>

(Drawings by
 Monika Fehre)

glands) alternating with the carpels. Carpels 2, appressed;

ovules 4-8 in each loculus; style well-developed, almost cla-

vate, abruptly expanded above into a disc-like clavuncle

bearing the conical stigma surmounted by 5 dense tufts of hair.

Fruit consisting of 2 slender follicles (mericarps); seeds

cylindric, grooved on one side, without a coma of hairs;

embryo straight, embedded in thick endosperm. 7 species

in Europe, North Africa, Western and Central Asia, in tem-

perate zone.

Lectotype, <u>Vinca</u> <u>major</u> L. designated in Britton & Brown,

<u>Illustr. Fl. N. U. S.</u> 2nd ed. 3:20 (1913); <u>V.</u> <u>minor</u> L. designated

by Hitchcock & Green in <u>Int. Bot. Congr. Cambridge, 1930,</u>

<u>Nomencl. Prop. Brit. Bot.</u> 136 (1929).

V. <u>CLAVIS</u>

Planta sempervirens, caulibus et foliis per hiemem persistenti-

      bus. Venae foliorum sub angulo 40°-50° e costa pro-

      deuntae:

    Calyx ca. 5-20 mm. longus:

      Sepala et folia margine glabra.......... 4. <u>V.</u> <u>difformis</u>

      Sepala et folia margine ciliata:

          Folia magna (aliquot plus quam 35 mm. longa) Planta

          florifera ad 30 cm. alta:

          Cilia calycis vix 0.2 mm. longa: Sardina

          .... 4. <u>V.</u> <u>difformis</u>
                                             ssp. <u>sardoa</u>

          Cilia calycis 0.5-1 mm. longa:... 6. <u>V.</u> <u>major</u>

        Folia parva (usque ad 35 mm. longa, 20 mm. lata).

          Planta florifera ca. 9-12 cm. alta: Peninsula

          Balcanica .... 5. <u>V.</u> <u>balcanica</u>

    Calyx ca. 3-5 mm. longus, semper glaber. 1. <u>V.</u> <u>minor</u>

Planta herbacea, caulibus et foliis hieme emorientibus:

    Venae foliorum sub angulo 5°-30° e costa prodeuntae:

      Folia utroque latere basi costae una vena tantum

vel venatione imperceptibili. Calyx longitudine

1/3-1/2 tubi corollae. Caules plerumque procum-

bentes vel prostrati, raro erectiusculi. Europa

orientalis, Asia occidentalis...... 2. <u>V</u>. <u>herbacea</u>

Folia utroque latere basi costae venis 3. Calyx

longitudine ca. 2/3 tubi corollae. Caules semper

erecti. Asia centralis .... 3. <u>V</u>. <u>erecta</u>

VI. <u>KEY TO SPECIES AND VARIANTS OF VINCA</u>

1. Plant evergreen. Veins of the leaves diverging from the

midrib at an angle of 40-50°........... <u>See</u> 2

- Plant dying down completely in winter. Veins of the

leaves diverging from the midrib at an angle of 5-30°

.... <u>See</u> 6

2. (1) Calyx about 5-20 mm. long. Leaves mostly broadest

below the middle .... <u>See</u> 3

- Calyx about 3-5 mm. long. Leaves mostly broadest at

the middle .... 1. <u>V</u>. <u>minor</u>
(<u>See</u> 13)

3. (2) Sepals and leaves completely without hairs

.... 4. <u>V</u>. <u>difformis</u>
ssp. <u>difformis</u>
(<u>See</u> 21)

- Sepals and leaves with hairs (visible under a lens)

on margin .... <u>See</u> 4

4. (3)   Sepals with very minute hairs (<u>ca</u>. 0.2 mm. long)

on margin, Sardinia .... 4. <u>V</u>. <u>difformis</u>
ssp. <u>sardoa</u>

-           Sepals with evident hairs (<u>ca</u>. 0.5-1 mm. long) on

margin .... <u>See</u> 5

5(4)   Flowering shoots to 30 cm high. Leaves some or

most more than 3.5 cm. long .... 6. <u>V</u>. <u>major</u>
(<u>See</u> 7)

-           Flowering shoots 9-12 cm. high. Leaves not more

than 3.5 cm. long. Balkan Peninsula

.... 5. <u>V</u>. <u>balcanica</u>

6(1)   Leaves with only one vein arising each side of mid-

rib at base or veining, not evident. Calyx about

1/3 to 1/2 length of corolla tube. Stems usually

prostrate or procumbent rarely somewhat erect.

Europe, Western Asia .... 2. <u>V</u>. <u>herbacea</u>

-           Leaves with 3 veins arising each side of midrib at

base. Calyx about 2/3 length of corolla tube.

Stems always erect. Central Asia .3. <u>V</u>. <u>erecta</u>

A. <u>Variants of V. major</u>

7. (5)   Petioles and young shoots almost glabrous, the hairs

few. Hairs of sepals to 0.5 mm. long. Europe

.... Subsp. <u>major</u>
(<u>See</u> 8)

-           Petioles and often the young shoots profusely hairy.

Hairs of sepals to 1.0 mm. long. Caucasus, north-

east Asia Minor .... Subsp. hirsuta

8. (5)   Corolla segments narrow (about 4 times as long as

broad), pointed, violet, 3-9 mm. broad; leaves

mostly lanceolate .... 'Oxyloba'

-      Corolla segments broader (nearly as broad as long),

blue or white, 10-20 mm. broad; leaves mostly

ovate .... See 9

9. (8)   Corolla white .... 'Alba'

Corolla blue .... See 10

10. (9)   Leaves plain green, not variegated ..... Typical major

Leaves variegated .... See 11

11. (10)   Leaves with network variegation .... 'Reticulata'

-      Leaves blotched .... See 12

12. (11)   Leaves with yellowish margins, darker green at

centre .... 'Variegata'

-      Leaves with dark green margin and lighter green

center .... 'Oxford'

B. Variants of V. minor

13. (2)   Leaves plain green, not variegated..... See 14

-      Leaves variegated................... See 19

14. (13)   Flowers single, i.e. corolla segments. See 15

-         Flowers double i. e. corolla segments about

          10-15 .... See 18

15. (14) Corolla white .... 'Alba'
                                                'Bowles's White'
                                                'Gertrude Jekyll'

-         Corolla blue, purple or reddish........ See 16

16. (15) Corolla blue .... See 17

-         Corolla red-purple.................. 'Atropurpurea'
                                               ('Azurea')

17. (16) Corolla segments to 9 mm. broad...... Typical minor L.

-         Corolla segments to 15 mm. broad..... 'La Grave'

18. (14) Corolla blue .... 'Plena'

          Corolla purple .... 'Multiplex'

          Corolla rose .... 'Roseoplena'

          Corolla white .... 'Alboplena'

19. (13) Leaves white-variegated .... 'Argenteo-
                                               variegata'

-         Leaves yellow-variegated............. See 20

20. (19) Corolla blue......................... 'Variegata'

-         Corolla white...................... 'Variegata-
                                              alba'

C. Variants of V. difformis subsp. difformis

21. (3) Corolla pale greyish blue............. Typical difformis

-         Corolla blue with a white eye........... 'Bicolor'

-         Corolla deep blue.................... 'Dubia'

1. VINCA MINOR L.

Vinca Pervinca Brunfels, Herb. Vivae Eicones 1:178, fig (1530)

Clematis daphnoides Fuchs, Hist. Stirp. 360, fig (1542), New

    Kreuterb. cap. 135, fig (1543).

Clematis daphnoides minor C. Bauhin, Pinax 301 (1623)

Pervinca vulgaris angustifolia Tournefort, Inst. Rei. Herb.

    120 (1700)

Vinca foliis ovatis α, β, γ, δ, L. , Hortus Cliff. 77 (1738)

Vinca minor L. , Sp. Pl. 1:209 (1753); A. DC. in DC. , Prodr.

    8:383 (1844); Wilkomm & Lange, Prodr. Fl. Hispan. 2:

    665 (1870); Coste Fl. France 2:546, f. 2469 (1902); Rouy

    Fl. France 10:225 (1908); Church, Types Fl. Mech. 201

    (1908); Fiori, Nuova Fl. Anal. d'Italia 2:245 (1925); Hegi,

    Fl. Mittel-Europa 5. iii:2053 fig. 3032, t. 228 f. 1 (1927)

    Hayek, Prodr. Fl. Penins. Balcan (Fedde, Repert Sp.

    Nov. , Beih. 30) 2:427 (1930); Ross-Craig, Draw. Brit. Pl.

    20:t. 24 (1964). Locus classicus: "Habitat in Germania,

    Anglia, Gallia" (Linnaeus, 1753). Type: Linnaean Herb.

    299. 1 (Linnean Soc. , London).

Pervinca minor (L.) Scopoli, Fl. Carniol. 2nd ed. 1:170 (1772)

    Caruel in Parlatore, Fl. Ital. 6:708 (1886); Williams, Prodr.

    Fl. Brit. 225 (1909).

Vinca humilis Salisbury, Prodr. Stirp. Chapel Allerton 146

 (1796), nom. illegit. superfl.

 Subshrub with trailing shoots up to 60 cm. long, rooting
at the nodes at intervals, making interwoven masses up to
20 cm. high, from which arise the short erect flowering
shoots. Leaves evergreen, the blade mostly lanceolate or
elliptic but often ovate on trailing shoots, with the base
rounded or cuneate, the apex acute or obtuse, 1.5-4.5 cm. long,
0.5-2.5 cm. broad, glabrous, pinnately veined; petiole 2-10 mm.
long. Flowering shoots erect, to 20 cm. long, with 1 or 2
axillary flowers; peduncles 1.5-2.5 cm. long. Calyx 3-4 mm.
long; segments narrowly lanceolate, glabrous. Corolla tube
ca. 9-11 mm. long; limb 2.5-3 cm. across, typically blue-violet
but varying to pale blue, reddish purple and white, the segments obliquely obovate, almost truncate, 10-15 mm. long,
6-15 mm. broad. Follicles 2-3 cm. long. Chromosome number 2n=46; (fide Finn, Pannochia-Laj, Rutland, Bowden, K.
Jones).

 General Distribution: Europe, from western France and
northern Spain eastward over southern and Central Europe to
the Caucasus, ascending to 1320 m in Graubünden, in Switzerland; long cultivated and naturalized in many places, notably
in the United States.

SPAIN: northern Spain south to Valencia; <u>cf.</u> H. M. Willkomm

    & J. C. Lange, <u>Prodr. Fl. Hispan.</u> 2:665 (1870)

FRANCE: spread over the whole country, although probably

    an introduction or escape from cultivation in some places;

    <u>cf.</u> H. J. Coste, <u>Fl. France</u> 2:546 (1902)

GERMANY: widespread, but in the north apparently introduced;

    <u>cf.</u> G. Hegi, <u>Fl. Mittel-Europa</u> 5. iii:205 (1927).

BELGIUM: in the Jura and Ardennes; <u>cf.</u> E. De Wildeman &

    T. Durand, <u>Prodr. Fl. Belge</u> 3:708 (1899).

POLAND: in lower mountain woods; <u>cf.</u> W. Szafer, S. Kul-

    czynski & B. Pawlowski, <u>Rosl Polskie</u> 608 (1953).

CZECHOSLOVAKIA: in Bohemia, Moravia and Slovakia; <u>cf.</u>

    J. Dostál, <u>Kvetena ČSR</u> 1150 (1950).

ROMANIA: widespread; <u>cf.</u> E. Nyárády in T. Savulescu, <u>Fl.</u>

    <u>Republ. Pop. Romine</u> 8:484 (1961)

AUSTRIA: widespread; <u>cf.</u> E. Janchen, <u>Cat. Fl. Austriae</u> 1:562

    (1958), G. Beck von Mannagetta, <u>Fl. Nieder-Österr.</u> 2:944

    (1893).

SWITZERLAND: widespread, ascending to 1200 m. in Valais

    (Wallis) and to 1320 m. in Graubünden; <u>cf.</u> P. Chenevard,

    <u>Cat. Pl. Vasc. Tessin</u> 383 (1910), J. Braun-Blanquet & E.

    Rübel, <u>Fl. Graubünden</u> 3:1124 (1934).

ITALY: spread over the whole country south to north-eastern
Sicily; cf. T. Caruel in F. Parlatore, Fl. Ital. 6:709 (1886),
A. Fiori, Nuova Fl. Anal. d'Italia 2:243 (1926).

JUGOSLAVIA: in the north; cf. J. F. Freyn, Fl. Süd-Istrien
139 (1877).

U.S.S.R.: natural occurrence uncertain, as in many places
it is apparently an escape from cultivation. In the Flora
SSSR 18:648 (1952), it is recorded from European Russia
as found in six floristic regions: Baltic States, Middle
Dnieper, Black Sea area, Bessarabia, Upper Dniester
Crimea, and in the Caucasus only in the northern part
of Western Transcaucasia. For its Caucasian localities,
cf. N. I. Kusnezow, Fl. Cauc. Crit. IV. 1:418 (1905), A. A.
Grossheim, Fl. Kavkaza, 2nd ed., 7:218, map 233 (1967).

Vinca minor is much more frost-tolerant than V. major
and has been esteemed as a garden plant for many centuries.
It is also more variable, producing variants with variegated
leaves and double flowers as well as single flowers in a diversity of colors. The double forms lack stamens, these having
become petaloid and colored like the normal corolla segments,
the total number of segments being further increased by their
frequent fission almost to the base; late in the season plants
with double flowers in spring may produce a few single

flowers.   The nomenclature of these forms is not wholly clear

owing to the lack of colored illustrations associated with

definite names.   The following list of cultivars may not cover

them all.   Caspar Bauhin in 1623 noted their color variation

and doubling:

> "flores caerulei sunt: quandoque candidi; rarius
> rubentes vel subpurpurascentes: interdum flos
> alium florem medium continet".

Cv. 'Alba'

Vinca minor alba Weston, Bot. Univ. 1:350 (1770); Sweet,

Hort. Brit. 2:274 (1827).

Corolla white, single, Albino forms are likely to arise

anywhere within the wide range of the species and there

are in cultivation several white-flowered forms of inde-

pendent origin, e. g. the common form, Bowles's form

and Gertrude Jekyll's form.

Cv. 'Alboplena'

Vinca minor alboplena Boom, Nederl. Dendrol. 371 (1949)

Corolla white, double.

Cv. 'Argenteo-variegata'

Vinca minor argenteo-variegata Weston, Bot. Univ. 1:351

(1770); Rehder, Man. Cult. Trees 2nd ed. 797 (1940).

Vinca major foliis argenteis Loddiges ex Loudon, Arbor.

Frut. Brit. 2:1256 (1838).

Leaves variegated with white; corolla blue.

Cv. 'Atropurpurea'

Vinca minor atropurpurea Sweet, Hortus Brit. 274 (1827);

Rehder, Man Cult. Trees, 2nd ed. 797 (1940).

Vinca minor flore puniceo Loddiges ex Loudon, Arbor.

Frut. Brit. 2:1256 (1838).

Vinca minor cuprea Döll, Fl. Grossherz. Baden 2:811

(1858).

Vinca minor punicea Bean, Trees & Shrubs 2:661 (1914).

Corolla white, single.

A relatively large-flowered form, with pink-flushed buds.

Cv. 'Gertrude Jekyll'

Corolla white, single.

A relatively small-flowered and small-leaved form.

Cv. 'La Grave'

Corolla segments up to 1. 5 cm. broad, often touching,

blue. This large-flowered form was introduced into cul-

tivation by Edward August Bowles, who discovered it

growing in the churchyard of La Grave on the road be-

tween Grenoble and the Col du Lautaret, Dauphiné,

France. Bowles told me in 1932, only a few years after

its introduction, that it covered the graves there and
that, although not in flower, its relatively broad foliage
induced him to collect it. He was delighted to find when
it flowered next year in his garden at Myddelton House,
Enfield, that it also had bigger broad-petalled flowers than
typical <u>V</u>. <u>minor</u>. This information I recorded in 1932 on
a specimen in the Cambridge University Herbarium (COE)
and also in my notes. The plant is also known as 'Bowles Variety'.

Cv. '<u>Multiplex</u>'

<u>Vinca</u> <u>minor</u> <u>rosea</u>, <u>V</u>. <u>minor</u> <u>atroviolacea</u>, <u>V</u>. <u>minor</u>
<u>cupricolor</u> Hegi, <u>Fl. Illustr. Fl. Mittel-Europa</u> 5. iii. 2054
(1927).

Whether there exist several forms of <u>Vinca</u> <u>minor</u> with
reddish purple flowers differing in intensity etc., or whether
there is only one form, which has received several names,
is uncertain. The only one I have seen has plum-purple
corollas (R. H. S., H. C. C. <u>9</u> 34 new ed. 79B) and, as noted
by Margery Fish in <u>Gard. Chron</u>. 161:19 (May 1967), is sometimes known in gardens as "Burgundy".

In the literature there are several possible epithets for
it, i. e. <u>atropurpurea</u> described by Sweet as 'dark purple',
<u>flore</u> <u>purpureo</u> described as "purpurrot", <u>cuprea</u> described

as "mit rothlich-kupferfarbenen Bluthen", _rosea_ described

as "Bluten violettrot", _atroviolacea_ described as "Bluten

schwarzviolet" and _cupricolor_ described as "Bluten kupfer-

farbigen". 'Atropurpurea' has been adopted here as being

the earliest available epithet.

Cv. 'Azurea'

>   Vinca minor azurea Dippel, Handb. Laubh. 1:157 (1889)

>   Corolla sky-blue, single.

Cv. 'Bowles's White'

>   Vinca minor multiplex Sweet, Hortus Brit. 2:274 (1827);

>   Rehder, Man. Cult. Trees 797 (1940).

>   V. minor flore pleno purpurea of gardens.

>   Corolla double, purple.

Cv. 'Plena'

>   Vinca minor plena Hubbard in Bailey, Stand. Cycl. Hort.

>   3:3471 (1928).

>   Corolla double, blue

Cv. 'Roseoplena'

>   Vinca minor rosea plena Dippel, Handb. Laubholzk. 1:157

>   (1889).

>   Vinca minor var. multiplex f. rosea Bergmans, Vaste

>   Pl. 2nd ed. 903 (1939).

Vinca minor roseoplena Boom, Nederl. Dendrol. 371

(1949).

Corolla double, violet-rose.

Cv.   'Superba'

Stated to be superior to 'La Grave' and described by

Messrs George Jackman as a

"dense and vigorous carpenter very free with
its blue flowers".

Cv.   'Variegata'

Vinca minor variegata Weston, Bot. Univ. 1:351 (1770);

Sweet, Hort. Brit. 274 (1827).

V. minor foliis aureis Loddiges ex Loudon, Arbor. Frut.

Brit. 2:1256 (1838).

V. minor aureo-variegata C.K. Schneider, Illustr. Handb.

Laubh. 2:849 (1912). Leaves with yellowish variegation.

Corolla blue.

Cv.   'Variegata-Alba'

Leaves with yellowish variegation.

Corolla white.

2.   VINCA HERBACEA Waldst. & Kit.

Vinca herbacea Waldstein & Kitaibel, Descr. Ic. Pl. Rar.

Hungar. 1:8, t. 9 (1799); Sims in Bot. Mag. 45:t. 2002 (1818);

Edwards, Bot. Reg. 4: t. 301 (1818); A. DC. in DC. , Prodr.

8:383 (1844); Reichenbach, Ic. Fl. Germ. 17: t. 1063 (1854);

Boissier, Fl. Orient. 4:45 (1875); Hegi, Fl. Mittel-Europa

5. iii:2052, fig. 3029 (1927); Hayek, Prodr. Fl. Penins.

Balcan. (Fedde, Repert. Sp. Nov. 30) 2:428 (1930); Jávorka

& Csapody, Magyar Fl. Képek. 403, f. 2756 (1932), t. 30,

f. 2756 (1933); Pichon in Bull. Mus. Nat. Hist. Nat. Paris II

23:440 (1951); Goulandris & Goulimis, Wild. Fl. Greece

84, t. (1968). Baytop in J. Fac. Pharm. Istanbul 7:16 (1971).

Locus classicus: "Crescit in montibus et clivis apricis

arenosis et calcareis Budae, Pestini, nec non in monti-

culis calcareis Comitatis Bihariensis" (Waldst. & Kit. ,

1799).

V. pumila E. D. Clarke, Travels 2. iii:555 (1816).

Locus classicus: Bulgaria, Mount Haemus,"in the route

between Constantinople and Rustchuk" (E. D. Clarke,

1816) .

V. libanotica Zuccarini in Abh. Akad. Wiss. Math. -Nat. München

3:246, t. 8 (1840); A. DC. in DC. , Prodr. 8:383 (1844);

Boissier, Fl. Orient. 4:46 (1875).

Locus classicus: Lebanon; "Crescit in subalpinis umbrosis,

v. g. in cedreto montis Libanon. Floret initio Majii (Iter

Schubert)"(Zucc. , 1840).

<u>V</u>. <u>herbacea</u> var. <u>pusilla</u> A. DC. , <u>Prodr</u>. 8:383 (1844).

    Locus classicus: "In campis apricis editoribus Bessarabiae"

    (A. DC. , 1844).

<u>V</u>. <u>herbacea</u> var. <u>glaberrima</u> A. DC. , <u>loc. cit</u>. (1844).

    Locus classicus: "In Zanto, Cilicia (Auch. 1503) et in

    Persia" (A. DC. , 1844).

<u>V</u>. <u>herbacea</u> var. <u>grandiflora</u> A. DC. , <u>loc. cit</u>. (1844).

    Locus classicus: "In sylvaticis montis Tactali Asiae

    minoris (Boiss. !) et prope Aleppum (Kotschy! 51)" (A.

    DC. , 1844)

<u>V</u>. <u>herbacea</u> var. <u>pusilla</u> A. DC. , loc. cit. (1844)

<u>V</u>. <u>sessilifolia</u> A. DC. in DC. , <u>Prodr</u>. 8:383 (1844).

    Locus classicus: "In Cappadocia ad Euphratem (Auch!

    N. 1498)" (A. DC. , 1844).

<u>V</u>. <u>herbacea</u> var. <u>gracilis</u> Griseb. , <u>Spicil. Fl. Rumel</u>. 2:66

    (1844).

    Locus classicus: "In Thracia: Sparsim gregarie in

    fruticetis Chersonesi alt. 900 pr. Ainadgik (substr. sax.

    aren.)" (Griseb. 1844).

<u>V</u>. <u>bottae</u> Jaubert & Spach, <u>Illustr. Fl. Orient</u>. 2:t. 186 (1847).

    Locus classicus: "In Mesopotamia legit cl. Botta! anno

    1845 (Herb. Mus. Par.)" (Jaub. & Spach, 1847).

<u>V</u>. <u>herbacea</u> var. <u>libanotica</u> (Zucc.) Kuntze in <u>Acta</u> <u>Horti</u>

Petrop. 11:210 (1887); Pichon in Bull. Mus. Nat. Hist. Nat.
Paris II, 23:446 (1951) p. p.

V. herbacea subsp. mixta Velenovsky, Fl. Bulg. 380 (1891).

V. mixta Velenovsky, Fl. Bulg. 646 (1891).

Locus classicus: Bulgaria; "In declivibus m. Balkan ad
Sofian (SK), Sliven (SK)" (Velenovsky, p. 380), "ad Sadovo
et . . . . . ad Trnovo" (Velenovsky, p. 646).

V. haussknechti Bornmüller & Sintenis ex Bornmüller in
Oesterr. Bot. Zeitschr. 48:453 (1898).

Locus classicus: Turkey; "In monte Deli-dagh inter
Siwas et Divriki 1893, vii (Bornm. iter Persico-turcicum
exs. no. 13436); Egin, Kota, in declivibus lapidosis 1890,
vi 6 (Sint. iter Orientale, exs. no. 2247)" (Bornmüller,
1898).

V. herbacea subsp. libanotica (Zucc.) Bornmüller in Beih.
Bot. Centralbl. 31. ii:239 (1914).

V. herbacea var. sessilifolia (A. DC.) Pichon in Bull. Mus.
Nat. Hist. Paris, II. 23:441 (1951).

Herbaceous perennial, dying back completely to the rootstock each winter, with trailing or ascending shoots to 20 cm.
long (when ascending) to 60 cm. long (when trailing). Leaves
herbaceous, the blade varying from elliptic (lowermost leaves)
to narrowly elliptic (uppermost leaves) or from ovate to lanceo-

late, with the base cuneate, the apex acute, 6-50 mm. long,
2-30 mm. broad, shortly ciliate or scabrid or smooth on the
margin, inconspicuously pinnately veined with the veins
diverging from the midrib at up to 30°; petiole 1-4 mm. long.
Flowers at intervals along the shoots; peduncles 1. 5-4 cm.
long. Calyx 3-10 mm. long; segments narrowly lanceolate or
linear, shortly ciliate or scabrid or smooth along the margin,
1/3 to 1/2 the length of the corolla tube. Corolla tube 10-20
mm. long; limb 2. 5-3. 5 cm. across, blue-violet, rarely
white, the segments obliquely narrowly obovate, 10-20 mm.
long, 3-8 mm. broad, acute. Follicles 2. 5-3. 5 cm. long.
Chromosome number 2n=46 (fide Finn, Bowden, K. Jones).
General Distribution: Central and eastern Europe and western
Asia, from Czechoslovakia, Hungary and Austria southward
over the Balkan Peninsula and to Asia Minor to Palestine and
Iraq and eastward over European Russia and the Caucasus
to northern Iran.

CZECHOSLOVAKIA: only in Slovakia; cf. J. Dostál, Květena
     ČSR 1149.

AUSTRIA: in Burgenland and Lower Austria (Niederösterreich),
     very scattered; cf. A. Neilreich, Fl. Nieder-Oesterr. 2:471
     (1859), G. Beck von Mannagetta, Fl. Nieder-Österr. 2:944
     (1893), E. Janchen, Cat. Fl. Austriae 1:562 (1958).

HUNGARY: widespread in the Hungarian plain, central Hungarian
mountains and lower Danube region; cf. S. Jávorka & V.
Csapody, Magyar Fl. Képek 403, f. 2756 (1932); S. Jávorka &
R. de Soó, Magyar Növ. Kézik. 1:484 (1951).

BULGARIA: widespread; cf. J. Velenovsky, Fl. Bulg. 380 (1891),
N. Stojanov, B. Stefanov & B. Kitanov, Fl. Bulg. 4th ed.
2:853 (1967).

ROMANIA: widespread and found in many localities; cf. E.
Nyárády in T. Savulescu, Fl. Republ. Pop. Romne. 8:
483, t. 92, f. 2 (1961).

JUGOSLAVIA: few records but probably widespread.

GREECE: widespread; cf. E. von Halácsy, Consp. Fl. Graecae
2:295 (1902), N. A. Goulandris & C. N. Goulimis, Wild. Fl.
Greece 84, (1968).

U.S.S.R.: widespread in scrub, steppes, mountain slopes
and chalk outcrops in European Russia and the Caucasus;
recorded in the Flora SSSR 18:650 (1952) from the Volga-
Don, Upper Dniester, Middler Dniester, Lower Don,
Bessarabia, Black Sea area, Crimea, Ciscaucasia, Dagi-
stan, Western, Eastern and Douthern Transcaucasia
floristic regions; cf. E. G. Pobedimova in Fl. SSSR 18:650
(1952), M. I. Kotov, Fl. URSP 8:263, fig. 61 (1957), A. A.
Grossheim, Fl. Kavkaza, 2nd ed. 7:218, map. 234 (1967).

EUROPEAN AND ASIATIC TURKEY: the specimens seen are
concisely listed below according to the grid squares and
Turkish vilayets shown on the map in P. H. Davis, Flora
of Turkey 1:2 (1965); its occurrence around Istanbul is detailed by A. Baytop in J. Fac. Pharm. Istanbul 7:18 (1971).

TURKEY IN EUROPE: A1, Tekirdag: 1967, Baytop & Atila
10827 (E), 10874 (E), 11010 (E).

TURKEY IN ASIA (ANATOLIA):

A4, Kastamonu: Manisadjian 979 (K)

A5, Samsun: 1000 m., 1965, Tobey 927 (E); 500 m., 1967,
Tobey 1812 (E); 1400 m., 1965, Tobey 993 (E).

A5, Amasya: 400-500 m., 1889, Bornmüller 352 (BM, K);
Manisadjian 409 (K); 1500 m., 1964, Tobey 643 (E); 900 m.,
1964, Tobey 643 A (E), 450 m., 1966, Tobey 643 B (E).

A7, Gumusane: 1600 m., 1894, Sintenis 5454 (BM, E, K);
1960, Stainton 8233 (E).

A8, Artvin: 500 m., Stainton 8185 (E).

A9, Erzurum: 1800 m., 1968, Barclay 708 (K).

B1, Izmir: 1877, Maw (K).

B4, Ankara: 1892, Bornmüller 3069 (BM, E); 850 m., 1929,
Bornmüller 14372 (BM, K); 900 m., 1933, Balls 215 (E,
K); Lindsay 15 (K); 950 m., Coode & Jones 113 (E); 1600 m.,
1959, Brown 1294 (K).

B6, Malatya: 1000 m. , 1957, Davis & Hedge 27695 (BM, E).

B7, Malatya: Sintenis 2218 (K)

B7, Sivas: Bornmüller 13436 (K)

B9, Van: 1899-1900, Maunsell (BM).

C2, Denizl: 1895, Whittall 446 (K).

C2-4, Antalya: 1936, Tengwall 285 (K); 1959, Hennipman &
      others 726   (K).

C4, Icel: 1300 m. , 1965, Coode & Jones 814 (E. K).

C5, Icel: 1900 m. , 1959, Hennipman & others 1258 (K)

C6, Maras: 550 m. , 1957, Davis & Hedge 27328 (E); 1300-
      1500 m. , 1957, Davis & Hedge 27396 (BM, E).

C6, Gaziantep: 1888, Sintenis 478 (E); 900 m. , 1934, Balls
      822 (BM, E).

C7, Urfa: 1855, Loftus 94 (BM)

SYRIA: apparently only in northwest: Aleppo (Halab), 1841
      Kotschy 51 (K); between Aleppo and Aintab (Gaziantep),
      480 m. , 1865, Haussknecht (K); Abou Douhour, 400 m. ,
      Haradjian 1011 (fide Rechinger 1960); S. of Homs, 1943,
      Davis 5596 (K).

LEBANON: Tripoli, 1919, Lamington & Yeo (K); Baalbek to
      Zebedan, 1878, Post 784 (BM): Bekaa, prope Baalbek,
      1150 m. , 1910, Bornmüller 12146 (BM): Kunetsse, 1400 m. ,
      1933, Meinertzhagen (BM); ridge above Zahleh, 1800 m. ,

Post 672 (BM); above Fakaya (Jabal Sannin), 1900 m.,

Townsend 640421 (K), in regione subalpina jugi Sanin,

16-1800 m., 1897, Bornmüller 1123 (K); Cedretum, 1900 m.,

1855 Kotschy 303 (BM, K); foothills, Hermon-Damascus

road, 1000 m., 1945, Norris (BM).

PALESTINE: Tiberias, 200 m., 1912, Dinsmore 3612 (K);

Esdraelon, Lowne (BM, K), 1942, Davis 4725 (K); Hadera,

plain of Sharon, 1906, Aaronsohn (fide Oppenheimer &

Evenari 1941); Beersheba to Gaza, 1911, Meyers & Dins-

more 3340 (K).

IRAQ: widespread in the mountain region of northern Iraq

and extending slightly into the foothills region; the fol-

lowing specimens are listed under the districts shown on

the physiographic map in E. Guest & A. Al-Rawi, Flora

of Iraq, vol. 1 (1966).

Jabal Sinjar district (MJS): above Balad Sinjar, 1100-1300

m., 1948, Gillett 11135 (K)

Rowanduz district (MRO): Salahaddin (Salah ad-Din),

Arbil liwa, 1300 m., 1951, Mooney 4328 (K); above Shaqlawa,

1958, Poore 644 (K).

Upper Jazira district: 30 km. E. of Karsi, 1964, Barkley

& Brahin 8004 (K); W. of Mosul, 1932, Uvarov, (BM).

Arbil district (FAR): 8 km. NW of Erbil (Arbil), 600 m.,

1948, Gillett & Rawi 10513 (K); 7 km. N of Chinarak (China-
ruk) to Rania, 570 m., 1959, Rawi, Nuri & Kass 28376 (K).

Aulaimaniya district (MSU): 31 km. N of Kirkuk, 740 m.,
1959, Rawi, Nuri & Kass 28016 (K); 8 km. of Chemalchemal
(Chamchamal), 900 m., 1948, Gillett & Rawi 10601 (K);
near Chemalchemal (Chamchamal) 1929, Rogers 0193
(BM, K), 1958, Poore 327 (K); Bazian Pass (Darband-i
Bazian), up to 900 m., 1932, Ludlow-Hewitt 1940 (K);
Gweija Dagh (Goizha), above Sulaimaniya, 1400 m., 1948,
Gillett & Rawi 10616 (K); Dokan (Dukan), 600 m., 1958,
Meade 133 (BM); 30 km. S. of Dokan, 1964, Barkley &
Haddad 7565 (K).

IRAN (PERSIA): fairly widespread in the western and central
Elburz Mountains of northern Iran, mostly in the Mazandaran
region: Mazandaren, distr. Nur. Kamarband, 2400-2600 m.,
1948, Rechinger 6418 (BM, K). In valle fluvii dalus 2200 m.,
1937, Rechinger 865 (BM, K). Elburz, 40 miles S of
Chalus, 1000 m., 1962, Furse 2524 (K). Mt. Kavadj.
Chalusi, 2700 m., 1963, Bowles Scholarship Exped. 751
(K). Central Elburz, 2100 m., 1963, Jacobs 6152 (K). 30
miles S of Amol, 900 m., 1964, Furse 5056 (K). 40 miles
N. of Tehran at Alikah Valiabad, 2700 m., Makinson 32
(K).

Vinca herbacea has the widest distribution and the greatest variability of any species of Vinca. Specimens vary in habit, in size, and in the shape of the leaf, in length and scabridity of calyx, in size and color of corolla, presenting many intergrading combinations of such characters with but little geographical correlation of them. Extreme specimens such as Manisadjan 979 with leaves only 12-20 mm. long and about 3 mm. broad and Bornmüller 13436 with leaves up to 80 mm. or more long and 35 mm. broad, are indeed very different, but intermediate specimens prohibit their separation as representatives of satisfactorily definable taxa. The area of greatest variability is in Iraq and eastern Turkey, where forms occur unequalled elsewhere in leaf size. The type-locality of V. herbacea is in Hungary. Here the plants have their leaves the broadest at the middle, in outline varying from elliptic to very narrowly elliptic, up to about 35 mm. long and 15 mm. broad, and calyx segments 4-5 mm. long with scabrid margins.

On Lebanon, the type-locality of V. libanotica, grow plants with leaves usually the broadest below the middle, in outline narrowly ovate or lanceolate, up to about 45 mm. long and 22 mm. broad, and calyx segments 5-7 mm. long with smooth margins. Remarkably broad leaves occur on specimens from eastern Turkey and Iraq. Examination of a large number of specimens reveals, however, no consistent correlation between such leaf and calyx characters.

Accordingly, V. herbacea, V. pumila, V. libanotica, V. sessilifolia, V. bottae, V. mixta and V. haussknechtii are here regarded as conspecific. It seems impossible to divide this complex into definable not intergrading taxa, but the general tendencies in variation can be summarized as follows: Leaves mostly broadest at the middle, elliptic to narrowly elliptic:

Calyx segments scabrid at the margin

Plants of Hungary, Austria, Romania, Greece

Calyx segments ciliate at the margin

Plants of Bulgaria, Ukraine, Crimea, Asia Minor.

Calyx segments glabrous and smooth at the margin

Plants of Caucasus and Iran

Leaves mostly broadest below the middle, ovate to lanceolate:

Calyx segments ciliate at the margin

Plants of Asia Minor

Calyx segments glabrous and smooth at the margin

Plants of Lebanon, Syria, Palestine, Asia Minor and

Iraq.

<u>V</u>. <u>herbacea</u> prefers more open habitats than other species,
growing in sunny places among rocks or on stony slopes, on the
edges of fields, in light scrub and on steppes. Such habitats
favor wider distribution than do woodlands, and <u>V</u>. <u>herbacea</u>
has thus attained a greater range and manifests a greater poly-
morphism than its congeners. The starry flowers vary much
in size; some forms, judging from herbarium material, would
be horticulturally more attractive than those at present in cultivation. In herbaria <u>V</u>. <u>minor</u> and <u>V</u>. <u>herbacea</u> have been con-
fused but may be distinguished <u>inter alia</u> by the shape of the
upper part of the corolla which is almost cylindric with almost
parallel sides in <u>V</u>. <u>herbacea</u> but obconic with outward diverging sides in <u>V</u>. <u>minor</u>.

3.   <u>VINCA ERECTA</u> Regel & Schmalh.

<u>Vinca erecta</u> Regel & Schmalhausen in <u>Acta Horti Petrop</u>. 6:

330 (1880); O. Fedtschenko & B. Fedtschenko, <u>Consp. Fl.</u>

<u>Turkest</u>. 5:13 (1913); B. Fedtschenko, <u>Rastit. Turkest</u>.

650 (1915); Pobedimova in <u>Fl.SSSR</u> 18:650, t. 35 f. 6. (1952).

Pavlov, <u>Fl. Kazakhst</u>, 7:123, t. 15 f. 1 (1964).

Locus classicus: Central Asia, Fergana valley:

"specimina fructifera legit Krause ad fluvium
Maili prope Samarkand"

(Regel & Schmalh., 1880).

Herbaceous perennial with several stems ascending from
a compact rootstock formed from short scaly stem-bases.
Leaves herbaceous, sessile, mostly narrowly ovate to elliptic, sometimes the lower ones ovate, with the base cuneate,
the apex acute, 2-5 cm. long, 1-3 cm. broad, glabrous or
pubescent on both surfaces, with 3 veins arising each side of
the midrib at the base and others near it, all almost parallel
and longitudinal. Flowering shoots erect, 15-40 cm. long,
with 1 or 2 axillary flowers; peduncles 3-5 cm. long. Calyx
7.5-15 mm. long; segments linear-lanceolate, glabrous or
pubescent. Corolla tube 10-20 mm. long; limb 3-4 cm.
across, pale blue or whitish, the segments lanceolate or elliptic to 15 mm. long, 5 mm. broad, acute. Follicles 3.5-5 cm.
long.

General Distribution: Central Asia, in mountains of
Kazakhstanskaya and Tadzhikskaya S.S.R. and adjacent north-
east Afghanistan.

U.S.S.R.: Kazakhstanskaya S.S.R. (Kazakhstan): in moun-
    tains east of Chimkent; cf. N.V. Pavlov, Fl. Kazakhst.
    7:123, t.15, f.1 (1964). Tadzhikskaya S.S.R. (Tadzhikistan):
    Prope Tutkaul, 1883, A. Regel (K). Ostia fl. Munzal in

fl. An-su, 1883, A. Regel (L.) Prope Usgent, 1200 m.,

1800 A. Regel (LE.)

AFGHANISTAN: Takhar province: unteres Farkhar-Tal,

Hange westlich von Farkhar, 1850 m., 1965, Podlech 10549

(M).

<u>Vinca erecta</u> is the most eastern and the most geographi-

cally isolated member of the genus but comes close to <u>V</u>.

<u>herbacea</u>, in which Pichon included it, and from which it has

presumably been derived. It is of erect habit, but erect

forms also occur in <u>V</u>. <u>herbacea</u>. The chief difference be-

tween the two is in the veining of the leaves, which is often

inconspicuous in <u>V</u>. <u>herbacea</u>, but marked in <u>V</u>. <u>erecta</u>.

In <u>V</u>. <u>herbacea</u>, only one distinct vein arises from each

side of the midrib at the base of the leaf, whereas in <u>V</u>. <u>erecta</u>,

there are three on each side, at the base. <u>V</u>. <u>erecta</u> varies

much in hair-covering. Boris Fedtschenko named, but did

not describe, a var. <u>glabra</u>, and a <u>hirsuta</u>.

I find no essential difference in the pubescence of the

inner supra staminal part of the corolla tube in specimens of

<u>V</u>. <u>erecta</u> from Tadzhikistan and <u>V</u>. <u>herbacea</u> from Hungary.

4. <u>VINCA DIFFORMIS</u> Pourret subsp. <u>DIFFORMIS</u>

<u>Vinca difformis</u> Pourret in Hist. Mém. Acad. Roy. Sci. Toulouse

3:337 (1788); Rouy, Fl. France 10:27 (1908); Stapf in

Curtis's Bot. Mag. 13:t. 8506 (1913); Sauvage & Vindt,

Fl. Maroc. 1 (in Trav. Inst. Sci. Chérif. Rabat, Bot. 4):120

(1952). Stearn in Bot. J. Linn. Soc. London 65:254, fig. 1E,

F (1972).

Locus classicus: south-west France, Pyrénées-Orientales;

"Frontfroide" (Pourret, 1788).

V. media Hoffmannsegg. & Link, Fl. Portug. 1:376, t. 70 (1813-

20); Wilkomm & Lange, Prodr. Fl. Hispan. 2:665 (1870);

Coste, Fl. France 2:546, fig. 2470 (1902).

Locus classicus: Portugal:

> "dans les haies, les lieux ombragés aux environs
> de Lisbonne et autre part"

(Hoffmannsegg. & Link, 1813-20).

V. acutiflora Bertoloni, Fl. Ital. 2:751 (1836).

Locus classicus: Italy:

> "Legi prope Sarzanam in collibus di Pellicione.
> Habui ex Liguria occidua a Lagueglia .....
> Interamna, et Roma ad sepes ..... ex Sardinia
> ..... Michelius in MS. cit. eam indicat Floren-
> tiae alle Cascine in initio nemoris"

(Bert. 1836)

V. major var. media (Hoffmgg. & Link) Ancibure & Prestat,

Cat. Pl. Rég. Bayonn. 41 (1918); Litardière in Briquet,

Prodr. Fl. Corse 3. ii:57 (1955).

V. major var. difformis (Pourret) Pichon in Bull. Mus. Nat.

Hist. Nat. Paris II. 23:442 (1951).

Subshrub with arching and trailing shoots. Leaves evergreen, the blade ovate to lanceolate, mostly narrowly ovate,
with the base rounded then abruptly or gradually attenuate into
the petiole, the apex obtuse or acute, 3-7 cm. long, 2.5-4.5
cm. broad, glabrous, pinnately veined; petiole 2-10 mm. long
with two minute glandular appendages above the middle.
Flowering shoots erect, to 30 cm. long, with 1-4 axillary
flowers at successive nodes; peduncles 1-4 cm. long. Calyx
ca. 6-12 mm. long; segments linear, glabrous, with a minute
gland each side a little above the base. Corolla tube ca. 6-
12 mm. long; limb 3-4 cm. across, pale blue or almost white,
the segments obliquely obovate, 12-20 mm. long, 7-13 mm.
broad, acute. Follicles to 3.5 cm. long. Chromosome number 2n=44 (fide K. Jones), 2n=46 (fide Bowden, Fernandes).

General Distribution: western Mediterranean region;
northwest Africa and southwest Europe from Morocco and
Algeria over Portugal, Spain and southern France to Corsica
and western Italy.

AZORES: recorded from S. Miguel, Terceira and S. Jorge
by R. T. Palhinha, Cat. Pl. vasc. Afores 93 (1966) but possibly introduced from Portugal and naturalized.

MOROCCO: recorded from the Tangier peninsula and the Rif

oriental et occidental, Maroc central (partie septentrionale),

Maroc occidental septentrional and Grand Atlas geographi-

cal divisions; cf. E. Jahandiez & R.C.J.E. Maire, Cat.

Pl.Maroc. 1:xi (1931), 3:580 (1934), C. Sauvage & J. Vindt,

Fl.Maroc. 1:120 (1952).

ALGERIA: common in the northern Mediterranean region: cf

J.A. Battandier & L. Trabut, Fl.Algérie, Dicot. 538

(1890); P. Quezel & S. Santa, Nouv.Fl.Algérie 2:745 (1963).

PORTUGAL: widespread and frequent; cf. A.X. Pereira

Coutinho, Fl. Portugal, 2nd ed. 577 (1939), C.A. de S.

Ferreira Sampaio, Fl.Portug. 2nd ed. 457 (1946).

SPAIN: widespread; recorded from Portevedia and Coruna

eastward to Tarragona and Barcelona, southward to

Cadiz, Malaga and Granada and from the Balearic Islands

(Majorca, Minorca); cf. H.M. Willkomm & J.M.C. Lange,

Prodr. Fl.Hispan. 2:665 (1870), H. Knoche, Fl. Balearica

2:290 (1922).

FRANCE: scattered in southern France from the Pyrénées -

Orientales (where it ascends to 1400 m.) over Hérault and

Var to Alpes-Maritimes and Corsica; cf. H.J. Coste, Fl.

France 2:546 (1903), G. Gautier, Fl.Pyrénées-Orientales

305 (1898), H. Loret & A. Barrandon Fl.Montpellier, 2nd

ed. 324 (1898), R.V. de Litardière in J.I. Briquet, Prodr.

<u>Fl. Corse</u> 3. ii. 57 (1955).

ITALY: in western Italy from Mentone southward to Portici

and Muro (Basilicata); <u>cf.</u> T. Caruel in F. Parlatore, <u>Fl.</u>

<u>Ital</u>. 6:709 (1886), A. Fiori, <u>Nuova Fl. d'Italia</u> 2:245 (1926).

<u>Vinca difformis</u> is closely allied to <u>V</u>. <u>major</u>, having the

same robust habit of growth, relatively large leaves and long

narrow calyx segments, but differs in being completely

glabrous. Its corolla is paler (near pale "Aster Violet", R.

H. S. H. C. C. 38/3), with proportionally narrower and more

pointed segments, thus having a star-like appearance. Under

cultivation in England, it is more liable to damage by frost

than <u>V</u>. <u>major</u>. These individually slight but collectively

significant differences are associated with a difference in

chromosome number, i. e. 2n=46 or 44 in <u>V</u>. <u>difformis</u> but

2n=92 in <u>V</u>. <u>major</u>. The ranges of the two overlap in southern

France and western Italy, but <u>V</u>. <u>difformis</u> occurs over a

large area, i. e. the Iberian Peninsula and northwest Africa

from which <u>V</u>. <u>major</u> is absent, or in which it is a sporadic

escape from cultivation.

Two color forms, one a deeper blue (<u>dubia</u>) the other with

a white centre, the rest blue (<u>bicolor</u>), have been distinguished,

and since their interest is primarily horticultural, may be

designated as cultivars:

Cv. 'Bicolor'

V. difformis β bicolor Coutinho, Fl.Portugal 485 (1913).

Described by Coutinho as having

"Limbo da corolla largaments branco no centro
e azul na parte restante. Arred. de Tavira".

Cv. 'Dubia'

V. media β dubia Battandie & Trabut, Fl.anal.Synop.

Algérie. 226 (1904). Described by Battandie and Trabut

as having

"fleurs très bleues, Babors, Tlemien"

and formerly in cultivation in Britain, having been intro-

duced by Professor Renée Maire.

VINCA DIFFORMIS subsp. SARDOA Stearn

"V. acutiflora" sensu Moris, Fl.Sardoa 3:66 (1859) non

Bertoloni (1836) sensu stricto.

"V. difformis" sensu Pannochia-Laj in Nuovo Giorn. Bot. Ital.

N.S. 45:158, fig. 31, 35, t. 14, f. 3a (1938) non Pourret.

V. difformis subsp. Sardoa Stearn in Bot.J. Linn.Soc. London

65:255, fig. 1 A-D (1972).

Suffrutex sempervirens, caulibus sterilibus reclinatis,

caulibus floriferis erectis 8-30 cm. longis e caulibus sterilibus

reclinatis, caulibus floriferis erectis 8-30 cm. longis e caudice vel e caulibus arcuatis exorientibus, omnibus glabris. <u>Folia</u>

variabilia, sempervirentia; lamina anguste ovata ad lanceo-

lata, basi in petiolum breviter attenuata, margine pilis

minutissimis vix 0.2 mm. longis scabra, superficie utrinque

glabra, pinnatim venosa; petiolus 5-12 mm. longus. <u>Flores</u>

<u>ca</u>. 4, axillares; pedunculi <u>ca</u>. 2-5 cm. longi, foliis breviores,

glabri. <u>Calyx</u> 5-12 mm. longus; segmenta peranguste tri-

angularia vel linearia, <u>ca</u>. 4.5-11 mm. longa, basi <u>ca</u>. 1 mm.

lata, margine denticulo minuto utrinque supra basin munita

et pilis minutissimis vix 0.2 mm. scabridula, apice penicillo

pilorum minutorum plerumque coronata. <u>Corolla</u> pallide

caerulea, magna; tubus totus anguste infundibuliformis, <u>ca</u>.

1.6 cm. longus, parte basali infrastaminali angusta <u>ca</u>. 5 mm.

longa, parte supera suprastaminali expansa <u>ca</u>. 10 mm. longa;

limbus stellatus, <u>ca</u>. 6-7 cm. diametro, segmentis oblique

obovatis, 2-4 cm. longis, 1-2 cm. latis, acutis. <u>Folliculi</u>

glabri. Chromosomata 2n=46 (<u>fide</u> Pannochia-Laj, 1938 , K.

Jones, 1972).

Subshrub with arching and trailing shoots. <u>Leaves</u> ever-

green, the blade mostly narrowly ovate to lanceolate, with the

base attenuate, the apex acute or acuminate, 2.5-7 cm. long,

1.5-4 cm. broad, glabrous except for the minutely ciliolate or

scabrid margin with hairs scarcely 0.2 mm. long, pinnately

veined; petiole 5-12 mm. long, glabrous, with two minute
glandular appendages above the middle. Flowering shoots
erect, 8-30 cm. long, with up to 4 axillary flowers at successive nodes; peduncles 2-5 cm. long. Calyx 5-12 mm. long;
segments very narrowly triangular or linear, minutely den-
ticulate and ciliolate along the margin with hairs scarcely
0. 2 mm. long and a tuft of short hairs at the apex and a minute
gland each side a little above the base. Corolla tube ca. 16 mm.
long; limb ca. 6-7 cm. across, pale, the segments obliquely
obovate, 20-40 mm. long, 10-20 mm. broad, acute. Follicles
glabrous. Chromosome number 2n=46 (fide Pannochia-Laj).
SARDINIA: Galura, prope Tempio Pausania, 500 m. , iii. 1912,
       Fiori in Fiori & Béguinot, Fl. Ital. Exsicc. II. 1727 (holo-
       typus, BM; K). Tempio, v. 1882, Reverchon, Pl. Sardaigne
       1882 n. 320 (BM, K). Cagliari, iii. 1887, Casella (BM)
    In the form of the corolla with acute segments broadest
near the middle this beautiful Sardinian periwinkle agrees with
ssp. difformis, but the corolla itself is somewhat larger with
the limb about 6-7 cm. across, instead of 3-5 cm. across,
as in ssp. difformis, and the upper part of the tube is a little
more widely expanded. The minute hairs along the slightly
scabrid margins of the calyx segments and the leaves also
distinguish it from ssp. difformis; they are shorter than those

of V. major, which has a differently shaped darker and smaller

corolla. Ssp. sardoa cannot satisfactorily be referred to either

V. difformis or V. major, but stands between them, except

in size of corolla. Its chromosome number, according to

Pannochia-Laj, who based her account on living plants sent

to Pisa from Sardinia, is 2n=46, as in Portuguese material

of true V. difformis, whereas V. major has 2n=92.

5. VINCA BALCANICA Pénzes

Vinca major var. pubescens Urumoff in Oester. Bot. Zeitschr.

    50:17 (1900); Chaytor & Turrill in Kew Bull. 1934: 441

    (1934); non (D'Urv.) Boiss. (1879).

    Locus classicus: Bulgaria, "a vicum Mikre", (Urumoff,

    1900), on northern slopes of Stara Planina . . . . .

V. balcanica Pénzes in Acta Bot. Acad. Sci. Hung. 8:329, fig. 1

    (1962).

    Locus classicus: "Albania: in inundatis rivi Terkura,

    prope pag. Tapiza" (A. Pénzes, 1962).

    Subshrub with procumbent sterile shoots and erect flowering

shoots. Leaves probably evergreen, the blade ovate or nar-

rowly ovate, with the base rounded, the apex acute, 1.2-3.5

cm. long, 0.5-2.0 cm. broad, ciliate on the margin with

hairs 0.6-1.0 mm. long, pinnately veined, with numerous

hairs along the micrib on the upper surface; petiole 3-7 mm.

long with minute glands above the middle. Flowering shoots

9-12 cm. high, one-flowered; peduncles 1-4. 5 cm. long.Calyx

5-9 mm. long; segments linear, sparsely ciliate. Corolla

tube 10-11 mm. long; limb ca. 2. 8 cm. across, blue the seg-

ments obliquely obovate, 16-19 mm. long, 11 mm. broad.

Follicles 1. 5 mm. or more long; chromosome number unknown.

General Distribution: Balkan Peninsula, in fairly adjacent

areas of Albania, Bulgaria and Jugoslavia.

BULGARIA: Mikre, 1899, Urumoff (fide Urumoff)

ALBANIA: near Tapiza, 1959, Karpati (fide Pénzes). On way

to Lushnija, 150 m., 1934, Pennington 128 (K). Vorra,

halfway Durazzo to Tirane, 300 ft. iv. 1936, Pennington

135 (K).

JUGOSLAVIA: between Prisren and Debra, 250 m. 1918 Kum-

merle (fide Pénzes). Zelinikova, 300-400 m., 1917 Born-

müller (fide Pénzes)

This little-known geographically isolated Balkan periwinkle

differs from V. major primarily in its weak and low growth

and much smaller leaves, characters which appear constant

according to the limited material available.

6. VINCA MAJOR L.

Clematis Daphnoides major C. Bauhin, Pinax 302 (1623)

Pervinca vulgaris latifolia Tournefort, Inst. Rei. Herb. 120,

    t. 46 (1700).

Vinca foliis ovatis ε L. , Hortus Cliff. 77 (1738).

Vinca major L. , Sp. Pl. 1:209 (1753); A. DC. in DC. , Prodr.

    8:384 (1844); Wilkomm & Lange, Prodr. Fl. Hispan. 2:666

    (1870); Coste, Fl. France 2:546, fig. 2471 (1902); Church,

    Types of Fl. Mech. 188 (1908); Hegi, Fl. Mittel-Europa

    5. iii; 2052, fig. 3030 (1927); Hayek, Prodr. Fl. Penins.

    Balcan. (Fedde, Repert. Sp. Nov. 30) 2:427 (1930).

    Locus classicus: "Habitat in Gallia Narbonensi, Hispania"

    (Linnaeus, 1753). Type: Linnaean Herb. 299. 3 (Linnean

    Soc. , London)

Pervinca major (L.) Scopoli, Fl. Carniol. 2nd ed. 1:170 (1772);

    Caruel in Parlatore, Fl. Ital. 6:710 (1886).

Vinca grandiflora Salisbury, Prodr. Stirp. Chapel Allerton

    146 (1796), nomen illegit. superfl.

Vinca ovatifolia Stokes, Bot. Mat. Med. 1:497 (1812), nomen

    illegit. superfl.

Vinca pubescens D'Urville in Mem. Soc. Linn. Paris I:282 (1822),

reimpr. in D'Urville, Enum. Pl. Ins. Arch. Pont.- Eux. 26
(1822).

Locus classicus: cf. infra (subsp. hirsuta)

Subshrub with arching and trailing shoots, the arching
shoots rising to about 30 cm. , then curving downwards and
rooting at their tips, the trailing stems rooting at intervals
and up to 100 cm. long. Leaves evergreen, the blade mostly
ovate or broadly ovate to lanceolate with the base usually
cordate but sometimes subcordate or truncate on trailing shoots,
the apex obtuse or acute, 2. 5-9 cm. long, 2-6 cm. broad,
ciliate on the margin with hairs 0. 1-0. 4 mm. (subsp. major)
or 0. 3-0. 8 mm. long (subsp. hirsuta), pinnately veined,
usually with numerous hairs along the midrib on the upper
surface; petiole 5-15 mm. long, almost glabrous (subsp. major)
to profusely hairy (subsp. hirsuta), with two minute glandular
appendages above the middle. Flowering shoots erect or
ascending, to 30 cm. high, with 1-4 axillary flowers at successive nodes; peduncles 1. 5-4 cm. long. Calyx ca. 7-18 mm.
long; segments linear, ciliate, with a minute gland a little above
the base. Corolla tube 12-15 mm. long; limb 3-5 cm. across,
typically blue-purple, sometimes white or violet, the segments
obliquely truncate, ca. 15-20 mm. long, 10-20 mm. broad, or
even lanceolate, ca. 15-20 mm. long, 3-10 mm. broad. Fol-

<u>licles</u> <u>ca</u>. 2.5-3.5 cm. long. Chromosome number 2n=92

(<u>fide</u> Pannocchia-Laj, Rutland, Bowden, K. Jones).

Subsp. <u>MAJOR</u>

<u>V</u>. <u>major</u> L., <u>Sp.Pl</u>. 1:209 (1753); <u>vide</u> <u>supra</u>.

<u>Pervinca</u> <u>major</u> (L.) Scopoli, <u>Fl.Carniol</u>. 2nd ed. 1:170 (1712);

    <u>vide</u> <u>supra</u>.

<u>V</u>. <u>grandiflora</u> Salisbury, <u>Prodr.Stirp.Chapel Allerton</u> 146

    (1796), nom. illegit.

<u>V</u>. <u>ovatifolia</u> Stokes, <u>Bot.Mat.Med</u>. 1:497 (1812), nom. illegit.

    <u>Leaves</u> mostly ovate, the marginal hairs 0.1-0.4 mm. long;

petioles sparsely hairy. <u>Calyx</u> segments ciliate with hairs to

0.5 mm. long. <u>Corolla</u> usually blue-violet, sometimes

white, the segments a little longer than broad, <u>ca</u>. 10-20 mm.

broad (in typical <u>V</u>. <u>major</u>) or much longer than broad, <u>ca</u>.

3-10 mm. broad (in cv. 'Oxyloba').

    General Distribution: undoubtedly native in southern

Europe, in southern France, Italy and northwest Jugoslavia,

but so long cultivated and so persistent a plant when naturalized

that its range independent of human activity can only be vaguely

surmised. In northern and central Europe it is an introduced

plant, as also in western Turkey, the Aegean Islands, Lebanon

and the United States, notably the Eastern and Pacific States.

PORTUGAL: recorded from one locality (Bucao, Beira), apparently introduced and naturalized; cf. A. X. Pereira

Coutinho, Fl. Portugal, 2nd ed. 577 (1939)

SPAIN: doubtfully native; recorded from a few scattered locali-

ties in Galicia, Andalucia, Valencia and Catalonia, where

it is likely to be derived from cultivation; cf. H. M. Wil-

komm & J. M. C. Lange, Prodr. Fl. Hispan. 2:666 (1870).

FRANCE: native in the Mediterranean region of southern

France from Pyrénées-Orientales over Hérault, Gard

and Var to Alpes-Maritimes, naturalized elsewhere, e. g.

in Auvergne and Corsica; cf. H. J. Coste, Fl. France 2:

546 (1903), G. C. C. Rouy, Fl. France 10:225 (1908), G.

Gautier, Cat. Fl. Pyrénées-Orientales 305 (1898), H. Loret

& Barrandon, Fl. Montpellier, 2nd ed. 324 (1886), J. B.

Saint-Lager Cat. Pl. Vasc. Bassin du Rhône 538 (1883),

Poulzolz, Fl. Dép. Gard. 2:43 (1862), A. Albert & E. Jahan-

diez, Cat. Pl. Vasc. Var. 332 (1908), H. J. P. Ardoino, Fl.

Alpes-Maritimes, 2nd ed. 258 (1879), R. V. de Chassagne,

Invent. Fl. d'Auvergne 2:360 (1957), Litardière in J. I.

Briquet, Prodr. Fl. Corse 3. ii:56 (1955).

ITALY: undoubtedly native and widespread from Piemonte and

Liguria eastward to Venezia-Friuli (cf. L. Gortani & M.

Gortani, Fl. Friulana 2:325 (1906) and southward over

Emilia, Toscana, Campania and Basilicata to Calabria

and Sicily, but in some places, particularly in north Italy,
naturalized from cultivation; cf. A. Fiori, Nuova Fl. Anal.
d'Italia 2:245 (1925), T. Caruel in F. Parlatore, Fl. Ital.
6:710 (1886).

JUGOSLAVIA: from Istria southward along the coastal region;
cf. A. von Degen, Fl. Velebitica 2:551 (1937).

GREECE: specimens have been collected from wild-growing
plants on the Ionian islands of Corfu (Kerkira), Cephalonia
(Kefallinia) and Zante (Zakinthos), in mainland Greece in
the Peloponnese (Peloponnisos), Crete and the Aegaean
islands of Milos, Chios (Khios), Ikaria, Samos and
Rhodes (Rodhos), but in most, if not all, of these localities it is likely to be derived from cultivation; cf. E. von
Halácsy, Consp. Fl. Graecae 2:294 (1902), K. H. Rechinger
f., Fl. Aegaea 558 (1943).

CENTRAL AND NORTHERN EUROPE: naturalized here and
there as an escape from cultivation; cf. G. Hegi, Illustr. Fl.
Mittel-Europa 5. iii:2052 (1927), A. R. Clapham, T. G. Tutin
& E. F. Warburg, Fl. Brit. Isles, 2nd ed. 639 (1962).

INDIA: naturalised in the Himalaya on the Singalila Range, West
Bengal (F. de Vos & E. G. Corbett 135) and in the Palni Hills,
Madras at 7500 ft. (W. Koelz 11262).

NORTH AFRICA: recorded from several places in Algeria but

apparently introduced, cf. P. Quézel & Santa, Nouv. Fl.

l'Algérie 2:745 (1963).

NEAR EAST: occasionally cultivated and sometimes naturalized

in western Turkey and Lebanon.

CULTIVARS OF V. MAJOR SUBSP. MAJOR

Like V. minor, which is, however, more variable, V.

major is represented in gardens not only by the typical wild

form with plain dark green leaves and blue-violet flowers,

but also by several forms which are best designated by culti-

var epithets as follows:

Cv. 'Alba'

Corolla white.

This form was listed in Booth and Son's nursery catalogue

for 1838 at 5s. per plant, for that time a very high price,

indicating its novelty.

Cv. 'Flavida'

Vinca major f. flavida Bergmans, Vaste Pl. 2nd ed. 903

(1939). Leaves at first yellowish, later green.

Cv. 'Multiplex'

Vinca major multiplex Sinclair in Donn, Hortus Cantabr.

12th ed. 102 (1831). Corolla double, blue.

Cv. '<u>Oxford</u>'

> Leaves with dark green margin and lighter green centre.
>
> Corolla blue. A form cultivated at the Oxford University
>
> Botanic Garden.

Cv. '<u>Oxyloba</u>'

> <u>Vinca</u> <u>major</u> var. <u>oxyloba</u> Stearn in <u>Gard. Chron.</u> III. 88:
>
> 156 (1930).
>
> "<u>V</u>. <u>major</u> subsp. <u>hirsuta</u>" sec. Stearn in <u>J. Bot.</u> (London)
>
> 70:Suppl. 27 (Aug. 1932) pro parte quoad plantam cultam;
>
> McClintock in <u>Proc. Bot. Soc. Brit. Isles</u> 5:341 (1964) pro
>
> parte quoad plantam in Britannia inquilinam; non (Boiss.)
>
> Stearn.

This plant has been cultivated in Britain for over 60 years
and has become naturalized here and there, as noted by Mc-
Clintock. It differs from typical <u>V</u>. <u>major</u> primarily in its
much narrower more pointed corolla lobes which are a deeper
violet (near "Campanula Violet" R. H. S., H. C. C. 37); the
whole corolla has a star-like appearance. The leaves are
predominantly lanceolate, rather than ovate. I described it
in 1930 as <u>V</u>. <u>major</u> var. <u>oxyloba</u> from a plant cultivated at
Myddleton House, Enfield, by E. A. Bowles, who had mentioned
it in his <u>My Garden in Autumn and Winter</u> 193 (1915) as "the

deep purple-blue, Italian plant, sometimes called acutiflora".
The same plant is illustrated in R. Gathorne-Hardy, The Tran-
quil Gardener 138 (1958) as V. major, and also in J. Nash,
English Garden Flowers 28 no. 4(1948) as "the puzzling plant".

Bowles knew nothing definite about its origin, but believed
it to be Italian. Later I found that it agreed in its narrow
corolla lobes and proportionally narrow leaves with Caucasian
specimens collected by G. Woronow in Abkhazia and I, accordingly, identified it as Vinca major subsp. hirsuta (V. pubescens).
Opportunity to consult more specimens from the Caucasus and
northeastern Asia Minor, and to compare living plants introduced to Kew in 1969 from the Caucasus, has convinced me
that they are not identical. The Caucasian plants are con-
stantly much more hairy than the cultivated oxyloba, particularly on the petiole, and have more widely spreading sepals
with longer hairs up to about 1 mm. long. Since, on the other
hand, there occur in southern Italy plants which, judging from
herbarium material, are virtually identical with cultivated
oxyloba (see below), rejection of its possible Caucasian origin
makes an Italian, indeed Sicilian origin, highly probable.

It is here treated as a cultivar pending an investigation of
Italian populations of V. major based on living material from
many localities, which has so far not been possible.

Cv. 'Reticulata'

Vinca major reticulata Donnaud, Nouv. Jardinier 1869:
1219 (1869); Bergmans, Vaste Pl. 2nd ed. 903 (1939).

Vinca major foliis reti culatis Dippel, Handb. Laubholzk.
1:158 (1889).   Leaves with yellow veining conspicuous
when young, later becoming green.   This is mentioned
in K. Koch, Dendrologie 2. i:291 (1872) as a variegated
form "die mit goldgelber Ädering sehr schön ist. "

Cv. 'Variegata'

Vinca major variegata Loudon, Hortus Brit. 67 (1830);
Loudon, Arb. Frut. Brit. 1254 (1838).

Vinca major elegantissima Donnaud, Nouv. Jardinier
1869:1219 (1869); Hibberd, Floral World 1871:224 (1871),
Hemsley, Handb. Hardy Trees & Shrubs 289 (1877);
Nicholson, Illustr. Dict. Gard. 4:160 (1887).   Leaves with
yellowish-white margin and dark-green centre.

Subsp. HIRSUTA (Boiss.) Stearn

Vinca pubescens D'Urville in Mem. Soc. Linn. Paris 1:282
(1822), reimpr. in D'Urville, Enum. Pl. Ins. Arch. Pont. -
Eux. 26 (1822); Pobedimova in Komarov, Fl. SSSR 18:
648 (1952); Grossheim, Fl. Kavkaza, 2nd ed. 7:218 map.
233 (1967).

Locus classicus: "In Colchide frequens, ad sepes"

(D'Urville, 1822), "Soukoum in Colchide" (D'Urville, 1822).

V. major var. hirsuta Boissier in Tchihatcheff, Asie, Mineure

III: Bot. 2:67 (1860).

Locus classicus: Asia Minor, Pontus: "collibus marit.

supra Samsun" (Tchihatchaff, 1860).

V. major var. pubescens (D'Urv.) Boissier, Fl. Orient.

4:45 (1879).

Vinca major subsp. hirsuta (Boiss.) Stearn in J. Bot. (London)

70:Suppl. 27 (Aug. 1932).

Leaves mostly lanceolate, the marginal hairs 0.3-0.8 mm.

long; petioles profusely hairy. Calyx segments ciliate with

hairs to 1.0 mm. long. Corolla with segments much longer

than broad, ca. 3-10 mm. broad.

General Distribution: northeastern Asia Minor and adja-

cent western Caucasus near the Black Sea.

TURKEY: vilayet Rize: vilayet Ordu: Cambas Road,

200 m., iii.1966, Tobey 1517 (E). Kayalar Pere, 20 m.,

iv. 1962, Tobey 50 (E). Environs de Rhize, Lazistan, v.

1866, Balansa (K). Cayeli east of Rhize, iv. 1959,

Guichard T. 39.59 (K).

U.S.S.R.: Abkhazia: Soukhoum, D'Urville (Geneva); prope

Suchum, iv. 1928, Gubbis in Grossheim & Schischkin,

Pl. Orient. Exsicc. 1928 n. 394 (BM, K); Sukhumi, 20 m.

vi. 1959, Davis 33660 (K). Georgia: inter Aczi et Bakhi,

Kutansskaya gub. , I. Kikodse, Iter Transc. 1914 (K). In

umbrosis circa Batumi, iv. 1917, Czerniawsky in Woronow,

Herb. Cauc. 1015 (K); Batumi, i. 1919, Rycroft (BM).

For garden purposes Vinca major, the greater periwinkle, is the most vigorous and ornamental member of the
genus. Its long flexible shoots with dark evergreen leaves
can easily be intertwined to make wreaths and garlands,
whence its classical name Vinca Pervinca. It grows readily
without care, survives neglect, and may even become naturalized; its large flowers are freely produced. In consequence
it has been spread widely by human activity during the past
two thousand years and now occurs in many places where it
cannot be native. However, even though its limits as a wild
plant have been blurred or obscured by past cultivation, it
would seem to be truly native in southern France and over
most of Italy. Typical major has large flowers with broad,
more or less truncate corolla segments, in color near
"Dauphins Violet", R. H.S. , H. C. C. o 39/1 or "Aster Violet"
A. C. C. 38/2.

In the Pontic region of the western Caucasus, and the
adjoining region of Asia Minor along the Black Sea coast,

occur plants distinguished from typical _major_ by more profuse
and longer hairs on the petiole and the margin of the leaf and
calyx and often on the stem, particularly at the nodes. Dumont
D'Urville gave the name _V. pubescens_ to such a plant occurring
at Sukhumi, Georgia, Caucasus in 1822 and Boissier the name
V. major var. _hirsuta_. to one occurring at Samsun, northern
Turkey. These two names are based on plants from near the
extremes of the above range. The Caucasian plant diverges
from typical _major_ of southwestern Europe, not only in its
greater pubescence, but also in its narrower more pointed
corolla segments and its proportionally narrower leaves. The
corolla difference is rendered less clear cut, however, by
specimens from Ordu, the Turkish vilayet east of Samsun,
which have broader, obtusely truncate, though well-separated
corolla segments. Nevertheless it does not seem possible to
regard the Pontic population comprising these extremes as
more than one taxon. This is certainly not identical with
typical _major_ and in 1932 I gave it subspecific rank, adopting
the earliest infraspecific epithet, i. e. _hirsuta_, and treating
V. pubescens and _V. major_ var. _oxyloba_ as synonymous.

The situation still remains puzzling, because the above
characters occur occasionally in Italian specimens. Thus a
specimen collected by Strobl in 1874 above Castelbuono,

Sicily (BM), has some leaves lanceolate, which, when young,

possess hairs almost as long as those of Caucasian specimens,

and also has pointed corolla segments only 4 mm. broad. The

abundance and length of the hairs on the calyx vary in Italian

material, but a specimen collected by Citarda in Boschi di

Valdemone (Todaro, Fl. Sicula Exs. 799; BM) is very like a

specimen collected by Tobey in vilayet Ordu, northern Turkey

(Tobey 50; E). The latter has glabrous carpels. The Sicilian

specimen has hairy carpels; the name V. major var. eriocarpa

Gussone, Fl. Sic. Syn. 1:281 (1843) was based on similar Sicilian

material with "petiolis, caulibus (plerumque) et folliculis

pilis longis albis conspersis". The name "Vinca (major β)

obliqua" was given by Pietro Porta in Nuovo Giorn. Bot. Ital. 11:

235 (1879) to a specimen collected in Calabria, southern Italy,

with apparently much the same kind of corolla, described as

with

> "corollae segmentis oblique truncatis, dextera parte
> ab apiculo quasi recta versus terminatis".

Fiori named V. major f. serotina a narrow-segmented form

from Campania. It is possible that two divergent populations

exist within Italian V. major, one of which approximates the

Pontic hirsuta.

Whatever interpretation is placed on these anomalous

taxonomic and phytogeographical facts, there remains undeniable
the close affinity of these Italian and Pontic populations now so
geographically isolated. The separation of these has been ef-
fected by past changes in vegetation and climate, and the distribution of land and water during the later Tertiary. Such conditions as these have left, for example, Narthecium reverchonii
in Corsica, N. scardicum in the Balkan Peninsula and N.
balansae (N. caucasicum) in the Caucasus. The occurrence,
in these Italian and Pontic populations of a few individuals
approximating to most individuals of the other population,
suggests that all their characters were present in the parental
stock of V. major. Thus, in the isolated Pontic population,
restricted to a small area, genetic drift has led to certain
characters becoming more general here than in the greater
western European population of V. major.

A very detailed and beautifully illustrated account of the
morphology of V. major will be found in Arthur Harry Church,
Types of Floral Mechanism (1908).

ACKNOWLEDGMENTS
This synopsis has been based primarily on specimens in
the herbaria of the British Museum (Natural History), London
(BM), the Royal Botanic Garden, Edinburgh (E) and the Royal

Botanic Gardens, Kew, (K), and on living plants cultivated at

the Royal Botanic Gardens, Kew, the Chelsea Physic Garden,

the Cambridge University Botanic Garden and the Royal

Horticultural Society's Gardens, Wisley.

Grateful acknowledgment is also to be made to the Komarov

Botanical Institute, Leningrad (LE), the Botanische Staats-

sammlung, Munich (M), and the Goulandris Botanical Museum,

Kifissia, Greece, for the loan of specimens, In general, only

Asiatic specimens have been listed as the European and North

Africa areas of the species are adequately covered by the

standard Floras cited.

In 1933 and 1934 I examined specimens of Vinca in the

herbaria of Geneva, Rome and Florence and some notes made

then have been used in the present account.

Special thanks are due to Dr. Keith Jones and his staff

at the Jodrell Laboratory, Kew, for chromosome counts of

plants of Vinca cultivated at Kew, and to Professor Abilio

Fernandes for a chromosome count of V. difformis at Coimbra.

ADDENDUM

2 bis.   VINCA SEMIDESERTORUM Ponert.

Vinca semidesertorum Ponert in Feddes Repert. 82: 581 (1971)

Locus classicus: U.S.S.R., Azerbaijan, 'inter oppidis Šamachi et Kirovabad' (Ponert, 1971).

Herbaceous perennial with shoots 20-30cm long. Leaves shor ciliate, the middle ones lanceolate-elliptic or lanceolate, 35-45mm long, 11-19mm broad, Peduncules 3-6cm long, longer than the leaves. Calyx 6-8mm long; segments shortly ciliate. Corolla tube about 15mm long, segments 15-20mm long, 6-8mm broad.

This is stated to be very close to V. herbacea and apparently comes within the variation of V. herbacea as here defined, although forming a semidesert population. Ponert's description was not received in England until mid-February 1972; this note has been inserted here simply to call attention to it.

REFERENCES

1. G.H.M. Lawrence, Baileya, 7, 113 (1959).

2. W.T. Stearn, Lloydia, 29, 196 (1966).

3. N.R. Farnsworth, Lloydia, 24, 105 (1961).

4. M. Pichon, Mém. Mus. Hist. Nat. XXXII, 6, 153 (1948).

CHAPTER 2

THE PHYTOCHEMISTRY OF VINCA SPECIES

Norman R. Farnsworth

Department of Pharmacognosy & Pharmacology
College of Pharmacy
University of Illinois
Chicago, Illinois

I. INTRODUCTION

Chemical and biological literature refer to six distinct

species of the genus Vinca, i.e., V. minor, V. major, V.

pubescens, V. difformis, V. herbaceae and V. erecta. How-

ever, according to Pichon,[1] there are only three Vinca

species, i.e., V. minor, V. major and V. herbacea, with

V. major existing as two simple varieties, var. major and

var. difformis, and V. herbacea existing as three simple

varieties, i.e., var. herbacea, var. libanotica and var.

sessilifolia. Regardless of which classification is most

correct, the former system will be utilized in this discussion

for the purpose of clarity and conformity to current literature.

In the past 15 years, at least 86 different alkaloids have been isolated from this genus of plants. Structure elucidation has been accomplished for all but 18 of these bases. The various Vinca species have each yielded the following number of alkaloids: V. minor (37), V. erecta (23), V. major (15), V. herbacea (13), V. difformis (9) and V. pubescens (3). Although our knowledge of the alkaloids from this genus is now rather complete, very little work has been accomplished with regard to non-alkaloid entities of the genus Vinca.

In addition to considering strict phytochemical information in this chapter, other minor aspects concerned with this group of plants will be briefly be mentioned in order to make our coverage of this genus as complete as possible. Lesser areas, such as folklore, distribution of species, biological activity of crude plant preparations, occurrence of Vinca alkaloids in other species of plants, and classical pharmacognostic (microscopic) studies reported in the literature will be included.

II.   THE PHYTOCHEMISTRY OF VINCA MINOR L.

A.   Description and Occurrence

Other botanical names that have been applied to V. minor, but none of them in the chemical literature, are Pervinca

<u>minor</u> (L.) Garsault, <u>Pervinca</u> <u>procumbens</u> Gilib., <u>Vinca</u>

<u>humilis</u> Salib. and <u>Vinca</u> <u>ellipticifolia</u> Stokes. Vernacular

names are petit pervenche, lesser periwinkle, evergreen

myrtle and myrtle. Equally common from western Europe to

Rumania, and in parts of european Russia, <u>V.</u> <u>minor</u> is fre-

quently found in the United States, where it is grown primarily

as a ground cover plant.[1-3]

This plant exhibits a dimorphic growth, with a sterile

creeping type of branch, and an erect flower-bearing type.

The leaves are elliptical, firm smooth, and have an entire

margin. The flower is typically blue, but white-flowering

variants exist, as well as horticultural forms.[1-5]

B. Folklore and Biological Properties

<u>Vinca</u> <u>minor</u> has been mentioned in the folklore as useful

for treating toothache,[6] snakebite,[6] hypertension,[7] and as a

carminative,[6] vomitive,[8] hemostatic,[8] and astringent.[6] It was

undoubtedly the folkloric use of this plant for hypertension

that prompted much of the recent phytochemical interest dis-

played in <u>V.</u> <u>minor.</u> In fact, with more than 35 alkaloids

having been isolated from this single species, it must be

considered as one of the most thoroughly studied plants in the

history of phytochemistry.

The biological effects of V. minor extracts have been reported by Hano and Maj[9, 10] in a series of studies using several species of animals. In addition, the action of galenical
preparations of this plant has been studied on the perfused
guinea pig heart.[11] The total alkaloids of V. minor were evaluated for a variety of pharmacologic effects by Quevauviller,
[12, 13] and Szczecklik et al.[7] treated several patients suffering
from arterial hypertension with powdered V. minor herb in
oral daily doses of 3 g. and concluded that the results obtained
were equivalent to those expected in patients treated for similar
conditions with reserpine, barbiturates or purines. It is interesting to note that Zheliazkov[14] has shown a dual action for V.
minor alkaloids. They not only produce a marked hypotensive
effect, but a curare-like action as well. It has been postulated
that the total alkaloids are composed of at least two general
types, the stronger type eliciting the hypotensive activity, and
the weaker type carrying the curare-like action.[14]

Aqueous extracts prepared from the whole plant of V.
minor have been shown to be toxic for American cockroaches
when injected into the bloodstream, but German cockroaches
and milkweed bugs were unaffected after immersion in the
extracts.[15]

An extract of V. minor leaves was inhibitory for Staphylococcus aureus, but inactive against Mycobacterium tuberculo-

sis, Escherichia coli and Salmonella typhimurium. Stem ex-

tracts were inactive against all four test organisms.[16] The

alkaloids of V. minor have never been evaluated for their anti-

microbial effects.

Buffered aqueous extracts of the leaves of two different

horticultural variants of V. minor gave 13 or 40 per cent re-

spective inhibitions of human plasma cholinesterase activity.[17]

Extracts of V. minor have been evaluated in a variety of

tumor systems and were found to be devoid of anticancer

activity.[18, 19]

Vincamine, the major alkaloid present in V. minor, is also

the most important one from a biological point of view. It is

used clinically in certain European countries, especially Hun-

gary, for the treatment of hypertension, angina and migraine

headaches.[20-27] The pharmacology of vincamine has been

studied in detail,[28-42] and these effects will be discussed in a

subsequent chapter.

C. Pharmacognostic Studies on Vinca minor L.

The microscopic characteristics of V. minor herb and

powdered V. minor have been recorded in detail by Blažek

and Starý.[43-45]

D. Non-alkaloid Constituents of Vinca minor L.

There have been no truly systematic studies on the determination of non-alkaloid constituents of V. minor, or in any other Vinca species. Those substances which have been isolated from this plant include dambonitol,[46] ornol,[47] fructose,[48] sucrose,[48] robinoside,[49] rubber,[50-51] 3-β-D-glucosyloxy-2-hydroxybenzoic acid,[48] β-sitosterol,[47] n-triacontane, and the ubiquitous triterpene ursolic acid.[47, 48, 50, 52-54] Other constituents that have been detected in V. minor, chiefly by means of paper chromatography, are listed in Table 1.

TABLE 1

Non-alkaloid Constituents Detected in Vinca minor

Constituent	Remarks
Caffeic acid[55]	In acid hydrolysates
Choline[56]	Doubtful presence
p-Coumaric acid[55]	In acid hydrolysates
Delphinidin-3, 5-diglycoside[57]	-
Gentisic acid[55]	In acid hydrolysates
p-Hydroxycoumaric acid[55]	In acid hydrolysates
o-Pyrocatechuic acid[55]	In acid hydrolysates
Protocatechuic acid[55]	In acid hydrolysates
o-Protocatechuic acid[55]	In acid hydrolysates
Tannins[48, 58, 59]	-
Vanillic acid[55]	In acid hydrolysates
Vincosides[58-60]	-

E. Alkaloids from Vinca minor L.

In 1950 Zabolotnaya reported on the isolation of an alkaloid from V. minor having the constitution $C_{22}H_{26}N_2O_3$, m.p.

223-224° (d) and $[\alpha]_D$ -8.4° ($CHCl_3$), which was named

minorine.[61] Later, in 1953, Schlittler and Furlenmeier iso-

lated a similar base from <u>V. minor</u> having the constitution

$C_{21}H_{26}N_2O_3$, m. p. 232-233° (d) and $[\alpha]_D$ +41°(pyr.), which

was named vincamine.[62] Čekan and co-workers, in 1959, es-

tablished that minorine and vincamine were indeed identical.[63]

Several workers have contributed to the structure elucidation

of vincamine (I),[64-71] and it has been synthesized by Kuehne.[72, 73]

      Schiendlin and Rubin[74] reported on the isolation of an alka-

loid from <u>V. minor</u> which they named perivincine, and which

was subsequently isolated from <u>V. major</u> by Farnsworth <u>et al.</u>

[75] In the hands of Trojánek, this alkaloid was found to be a

mixture of vincamine and vincine.[76, 77]

      Similarly, Szasz and co-workers,[78] as well as Trojánek <u>et</u>

<u>al.</u>,[54] and Lyapunova and Birosyak[79] isolated an alkaloid from

<u>V. minor</u> which was given the name isovincamine, but it was

also shown to be a mixture.[77]

      A perplexing question that exists with regard to the alka-

loid composition of <u>V. minor</u> concerns a report on the pres-

ence of reserpine. Lyapunova and Birosyak[80] reported on the

isolation of crystalline reserpine in a yield of 0.003% from the

air-dried roots of <u>V. minor</u> from the Ukraine. Their identifi-

cation was made on the basis of a melting point, mixture melting point, color reactions typical for reserpine (as well as for similar alkaloids), and chromatographic comparisons with an authentic sample.

Perhaps the failure of other workers, who have more extensively investigated V. minor, to confirm the presence of reserpine, is that they have used either leaf material, or aerial parts of V. minor, and not the roots. Another possibility is that the Ukrainian plant material is chemically different from that found in other european countries.

All of the remaining work on the alkaloids of V. minor, has been accomplished by four groups in Europe, namely Trojánek et al., Mokrý et al., Plat et al. and Döpke et al.

Trojánek et al. isolated the alkaloids vincaminorine,[54] vincaminoreine,[81,82] vincamidine (strictamine),[81,82] vincaminine,[83,84] vincine,[77] and deduced the structures for vincine (III),[85,87] vincinine (VI),[88] vincaminine (V),[88] vincaminorine (XIII),[89] and vincaminoreine (XIV).[89] These isolation studies were performed mainly by extracting the total crude bases in the usual manner, followed by column chromatography, pH extractions and preparative paper chromatography.

Mokrý et al. isolated the alkaloid vincanorine, which was shown to be identical with (+)-eburnamonine (VII_{dl})[90,91] and separated (-)-eburnamonine (VII_1) from the racemate,[89] in addition to the alkaloid vincareine,[90] which was identical with the

vincinine of Trojánek.[83, 84] They also isolated vincadine (XVI),

[92-94] (±)-minovine (XX),[92, 95] vincorine,[92] vincaminoridine(XV),

[96] vincoridine (LXIII),[97] (±)-Ind-N-methylquebrachamine (XVII),

[98] (±)-vincadifformine (XIX$_{dl}$),[95] (-)-vincadifformine (XIX$_1$),[95]

14-epivincamine (II),[99] 20-hydroxyvincamine (IV),[100]

eburnamenine (XII),[101] (+)-1, 2-dehydroaspidospermidine (XXX),

[101] (-)-eburnamine (pleiocarpinidine)(X),[101] (+)-isoeburnamine

(XI),[101] vinoxine,[101] (+)-N-methylaspidospermidine (XXIX)[101]

and (+)-quebrachamine (XVIII).[100] The alkaloid (+)-quebracha-

mine (XVIII) has been suggested to be an artifact produced by

the facile removal of the carbomethoxy group of vincadine (XVI)

during the extraction procedure.[100] Recently, the first two ex-

amples of dimeric indole alkaloids have been reported from V.

minor by Kompiš and Grossmann, namely vincalutine, a

$C_{39}H_{49}N_3O_8$ compound, and vincarubine, a $C_{45}H_{54}N_4O_7$

compound.

Mokrý et al. determined the structures for those alkaloids

indicated above that were isolated by them and were unknown

at the time of discovery, except for vincorine, vincalutine,

vincarubine and vinoxine, whose structures are still unknown.

The isolation studies of Mokrý et al. were based primarily

on an initial separation of the crude alkaloids by counter-current distribution, followed by column chromatographic separations.

Plat and co-workers ahve isolated and deduced the structures for minovincine (XXI), [103-105] 16-methoxyminovincine (XXII), [103-105] minovincinine (XXIII), [103-105] and have identified (-)-vincadifformine (XIX$_1$) [103-105] in V. minor. Minovincine is identical with Trojȧnek's minoricine. [85, 103, 105]

The latest workers in the field of V. minor alkaloids have been Döpke and co-workers, who isolated vincatine, [106] vincesine, [106] 11, 12-dimethoxyeburnamonine (IX) [107] and 11-methoxyeburnamonine (VIII). [108] Vincatine and vincesine are alkaloids of unknown structure. [106]

Several methods for the paper chromatographic separation of V. minor alkaloids have been suggested, [77, 78, 79, 81, 109-112] as well as a microassay of vincamine employing thin-layer chromatography. [113]

The physical data for all alkaloids isolated from V. minor are presented in Table 2, and the structures for all alkaloids from this plant, where known, are shown in Figures 1-16.

F.   Distribution of Vinca minor Alkaloids

Several of the alkaloids from V. minor have been isolated from related apocynaceous plants, particularly those bases of the eburnamine and quebrachamine types. Thus, eburnamenine

has been found in Aspidosperma quebracho-blanco,[114]Hunteria

eburnea,[115, 116] and Pleiocarpa mutica[117] and Rhazya stricta

[114]; eburnamonine of unknown optical rotation from Rhazya

stricta[114] and Amsonia tabernaemontana[118] and (+)-eburnamo-

nine from Hunteria eburnea[115]; (-)-eburnamine (pleiocarpini-

dine) in Hunteria eburnea,[115, 116] Pleiocarpa mutica,[116]

Pleiocarpa pycnantha,[119] Rhazya stricta,[114] Gonioma kamassi,

[120] Amsonia tabernaemontana,[118] and Haplophyton cimicidum,

[116] and (+)-isoeburnamine has been found in Hunteria eburnea,

[116] Amsonia tabernaemontana,[118] and Haplophyton cimicidum.

[116] Aspidosperma quebracho-blanco,[114, 122] Gonioma

kamassi,[120] and Rhazya stricta[114] have yielded (+)-1, 2-dehy-

droaspidospermidine, whereas (-)-1, 2-dehydroaspidospermi-

dine has been isolated from Pleiocarpa tubacina.[121] Stemma-

denia-donnell-smithii,[124] Pleiocarpa tubacina,[121] and Gonioma

kamassi[120] have yielded (+)-quebrachamine. On the other hand,

(-)-quebrachamine has been isolated from Melodinus australis,

[125] Aspidosperma sandwithianum,[126] Rhazya stricta,[114,123,133]
Aspidosperma quebracho-blanco,[122, 127] Aspidosperma

peroba,[128] Aspidosperma album,[128] Aspidosperma chakensis,

[129] Aspidosperma polyneuron[130-131] and Gonioma kamassi.[132]

V. minor has yielded (±)-Ind-N-methylquebrachamine,[98]

TABLE 2

Physical Data for Alkaloids Isolated from Vinca minor L.

Alkaloid	Structure[a]	Formula	m.p.(°C)	$[\alpha]_D$	λ_{max}(mμ)
Eburnamenine[101]	XII	$C_{19}H_{22}N_2$	amorph. 195 (pic.)	$+183^d$	225, 260, 301, 311
(±)-Eburnamonine[91,101]	VII$_{dl}$	$C_{19}H_{22}N_2O$	198-200 201-202.5	$\pm O^{cd}$ $\pm O^{de}$	-
(-)-Eburnamonine[91,106]	VII$_l$	$C_{19}H_{22}N_2O$	171.5 174-176	-85^d -90.2^d	-
(+)-1,2-Dehydro-aspidospermidine[101]	XXX	$C_{19}H_{24}N_2$	oil	$+226.7^d$	224, 257, 229 (infl.)
(-)-Eburnamine (pleiocarpinidine)[101]	X	$C_{19}H_{24}N_2O$	179-181	-98.2^d	230, 284
(+)-Isoeburnamine[101]	XI	$C_{19}H_{24}N_2O$	191-192	$+91.4^d$	229, 283
(+)-Quebrachamine[101]	XVIII	$C_{19}H_{26}N_2$	145-146	$+120^{cg}$	-
Vincamidine[81,101] (strictamine)	XLVII	$C_{20}H_{22}N_2O_2$	135 78-80^g	$+103^{dg}$ -	262, 290(sh.)

11-Methoxyeburnamonine[108]	VIII	$C_{20}H_{24}N_2O_2$	169-170	-107^d	247, 279
Vinoxine[101]	-	$C_{20}H_{24}N_2O_3$	216-218(d) (HCl)	-25^d	220, 270, 280, 290(sh)
(±)-Ind-N-methyl-quebrachamine[98]	XVII	$C_{20}H_{28}N_2$	70-72	$\pm O^c$	232, 289, 296
(+)-N-methyl-aspidospermidine[101]	XXIX	$C_{20}H_{28}N_2$	amorph.	$+24.4^d$	210, 258, 307
Vincoridine[96, 97]	LXIII	$C_{21}H_{24}N_2O_3$	159-160	-158^d	210, 258, 312
Minovincine[85, 103, 104] (minoricine)	XXI	$C_{21}H_{24}N_2O_3$	amorph. 192(HCl) 213-216(pic)	-504^c -534^b	230, 300, 328 228, 302, 328
Vincaminine[83, 84, 88, 90] (vincareine)	V	$C_{21}H_{24}N_2O_4$	205-206 208-210(d)	-28.2^d $+29.5^e$	227, 250, 286(infl.) 226, 275, 287
11,12-Dimethoxy-eburnamonine[107]	IX	$C_{21}H_{25}N_2O_3$	220	-	244, 272, 318
(±)-Vincadifformine[95]	XIX$_{dl}$	$C_{21}H_{26}N_2O_2$	125	$\pm O^{cd}$	-
(-)-Vincadifformine[95, 103]	XIX$_l$	$C_{21}H_{26}N_2O_2$	-	-540^c	228, 300, 330

Compound		Formula	M.P.	$[\alpha]$	UV
Minovincinine[103,104]	XXIII	$C_{21}H_{26}N_2O_3$	amorph.	-418^c	225, 297, 328
Vincamine[54,61-70,77]	I	$C_{21}H_{26}N_2O_3$	232	$+42^e$	-
			229-232	$+45$	
			230-232	-	
			232-233.5	$+40^e$	
			233-234	$+38^e$	
				-15^d	
			232-233	$+41^e$	228, 280
Isovincamine[54,78,f]	-	$C_{21}H_{26}N_2O_3$	215-217	$+37^e$	-
			218-220(d)	$+26^e$	
14-Epivincamine[99,101]	II	$C_{21}H_{26}N_2O_3$	189-191	-38.9^d	226, 276
			181-185	-36.4^d	
20-Hydroxyvincamine[100]	IV	$C_{21}H_{26}N_2O_4$	222-225	-	-
Vincadine[92-94,98]	XVI	$C_{21}H_{28}N_2O_2$	70-75	$+91.5^c$	228, 286, 290(infl.)
			78-82	$+86^d$	
16-Methoxy-minovincine (minoriceine)[85,103,104]	XXII	$C_{22}H_{26}N_2O_4$	amorph.	-414^c	230, 250, 325
			141-142	-554^b	248, 327

Compound	No.	Formula	M.p.	[α]	UV λmax
Vincinine [84,88]	VI	$C_{22}H_{26}N_2O_5$	202-204(d)	+ 24[e]	228, 272, 298 / 225, 266, 295
(±)-Minovine [85,92,95,103]	XX	$C_{22}H_{28}N_2O_2$	120-122 / 80-82 / 79-81	± O[cd] / ± O[b] / ± O[c] / ± O[c]	- / - / 338, 310(infl.) / 223, 308, 335
Vincorine [92]	-	$C_{22}H_{28}N_2O_3$	93-94	- 142[c]	254, 326
Vincine [77,86]	III	$C_{22}H_{28}N_2O_4$	212-214	+ 36[e]	222, 272, 296
Vincaminoreine [81,89,93]	XIV	$C_{22}H_{30}N_2O_2$	138-139 / 136-138	+ 26,5[d] / + 27[b]	230, 288, 296
Vincaminorine [54,89,94]	XIII	$C_{22}H_{30}N_2O_2$	130-131	+ 46[c]	226, 281, 291(infl.)
Vincatine [106]	-	$C_{22}H_{30}N_2O_3$	111-112	-	-
Vincesine [106]	-	$C_{22}H_{32}N_2O_4$	amorph.	- 476[c]	299, 327
Perivincine [74,f]	-	$C_{23}H_{26}N_2O_4$	201-202(d)	-	229, 274
Vincaminoridine [96,97]	XV	$C_{23}H_{32}N_2O_3$	99-100	+ 57.7[d]	232, 300, 288(infl.)
Reserpine [79]	LI	$C_{33}H_{40}N_2O_9$	262-263(d)	-	-

Vincalutine[102]	-	-	$C_{39}H_{49}N_3O_8$	amorph.	-464^c	224, 280, 415
Vincarubine[102]	-	-	$C_{45}H_{54}N_4O_7$	amorph.	-558^c	270, 336, 485, 288(infl)

[a]For structures see Figures 1-16, [b]Methanol, [c]Ethanol, [d]Chloroform, [e]Pyridine, [f]Shown to be a mixture of vincamine and vincine[76,77], [g]As hydrate.

whereas the (-)-isomer has been isolated from Vallesia

dichotoma,[134] Aspidosperma quebracho-blanco[122] has been

found to contain (+)-N-methylaspidospermidine, and

(+)-vincadifformine has been isolated from Tabernaemontana

riedelii,[135] Vallesia dichotoma,[134] Rhazya stricta,[123] and

Vinca difformis.[260, 261, 265] (-)-vincadifformine has been

separated from its racemate obtained from V. minor,[95, 103, 104]

and (+)-vincadifformine has been isolated from Rhazya

stricta[123] and Tabernaemontana riedelii.[135] Vincamine, the

major alkaloid of V. minor, occurs also in V. major,[233] V.

difformis,[139] and in V. erecta[140-142] in the form of its (+)

isomer, and (±)-vincamine, as well as (+)-vincamine have

been isolated from Tabernaemontana rigida.[135] Vincine has

been isolated only from V. minor[77, 86, 87] and V. major,[233]

and 14-epivincamine from V. minor[70, 99, 101] and Tabernae-

montana rigida.[135] (-)-Minovincinine has been isolated from

V. minor,[103, 104] and (+)-minovincinine from Alstonia venenata.[143] Vinca minor[103, 104] and Tabernaemontana riedelii[135] have

both yielded (+)-minovincine.

Reserpine has been isolated from several Rauvolfia species,[144-185] from Tonduzia longifolia,[186] Vallesia dichotoma,[187]

Catharanthus roseus (Vinca rosea),[188-190] Alstonia constricta,[191-193] Alstonia venenata,[194, 195] Excavatia coccinea,[196]

Ochrosia poweri,[196] and Aspidosperma auriculatum.[253] The

evidence for the presence of reserpine in C. roseus must be

considered as equivocal,[188-190] as must its presence in V.

minor.[80]

Thus, if we discount reserpine as being present in V.

minor, alkaloids from this plant occur in 14 species of nine

genera of the Apocynaceae, i.e. Aspidosperma, Haplophyton,

Pleiocarpa, Rhazya, Stemmadenia, Tabernaemontana,

Gonioma and Vinca.

III. THE PHYTOCHEMISTRY OF VINCA MAJOR L.

A. Description and Occurrence

V. major has a curiously disjunct distribution in France,

Italy, Switzerland, Sicily, Madeira, Canary Islands, Crete,

Rhodes, Chios, South and East shores of the Black Sea, and

probably introduced to England, Portugal, Spain, Malta,

Algeria and the United States.[1-3] It exhibits a dimorphic

growth; sterile creeping, and erect flowering branches. The

leaf is soft and obovate, and differs from the closely related

V. difformis Pourr. in having well ciliated margins of the

leaves and sepals.[1-3]

Other botanical names that have been applied to this

species include Pervinca major (L.) Garsault, Vinca grandiflora

Salisb., Vinca media Hoffmgg. et Link, Vinca ovatifolia Stokes, Vinca acutiflora Bertol., Vinca intermedia Tausch, Vinca obliqua Porta, Pervinca media (Hoffmgg. et Link) Caruel, Vinca obtusiflora Pau, Vinca lusitanica Brot and Vinca pubescens Urv.[1-3]

Pichon considers V. major L. to be Vinca major (L.) var. major Pich., but such nomenclature has never been applied to this plant in phytochemical investigations.[1] Also, Pichon considers V. pubescens Urv. to be synonymous with V. major (L.) var. major Pich.,[1] but this is disputed by Pobedimova,[197] who considers V. pubescens Urv. as an independent species and defines its morphological characters and other features, including its area of distribution, which appears to clearly distinguish it from V. major L.

Vernacular names applied to V. major L. are pervenche grande and the greater periwinkle.

B. Folklore and Biological Properties

V. major whole plant extracts have been reported to be used in France for abortifacient,[198, 199] antigalactagogue,[200, 201] and antihemorrhagic[201] effects, and as a tonic,[201] bitter,[201] and vulnerary,[201] and in the United States as a blood tonic.[202]

Leaf extracts have been employed in South Africa and in France as an astringent,[109, 201, 203] and in treating menorrhagia.[199, 203, 204] It is interesting to note that the leaf of V. major is reported used for diabetes in Natal,[199] but the plant is not known to grow in this region. Presumably, this reference should have been to Catharanthus roseus (Vinca rosea), which grows in Natal, and has a wide reputation as an antidiabetic remedy.[205]

V. major (V. pubescens Urv.) total alkaloids have been reported by Orechoff et al.[206] to elicit marked hypotensive effects in animals, and Quevauviller et al.[207] have reported on the pharmacological properties of V. major total alkaloids in dogs, and found that doses of 10 mg/Kg (iv) produced a marked fall in blood pressure.

A crude extract of V. major was devoid of antibacterial activity when tested against Staphylococcus aureus, Escherichia coli, Salmonella typhimurium, and Mycobacterium tuberculosis,[208] and extracts were similarly devoid of antitumor activity against a variety of tumor systems in animals.[209] The total alkaloids of V. major failed to induce a leukopenia in rats.[210, 211]

Vinine (carapanaubine)[206] has been shown to elicit hypotensive effects in animals, whereas reserpinine[212] and perivincine (vincamine + vincine)[75] gave only transient hypo-

tensive effects. Vincamine, also present in other Vinca species,

has been mentioned previously regarding its pharmacologic

effects in animals[28-42] and in humans.[20-27]

C. Pharmacognostic Studies on Vinca major L.

Blažek and Starý have reported on a detailed microscopic

examination of the cellular elements and structures of V.

major.[44]

D. Non-alkaloid Constituents of Vinca major L.

As with other Vinca species, there has been very little

interest in the non-alkaloidal entities of V. major. Wall and

co-workers[213] have screened extracts of V. major and have

reported a positive hemolysis test, probably due to triterpenoid

saponins, a positive test for the presence of sterols (Lieber-

man-Burchard), a positive test for organic acids, and negative

tests for flavonols (magnesium-HCl), tannins, phenols, and

alkaloids (?). Bate-Smith reported strong positive leucoantho-

cyanin tests on V. major.[214] Carotene[215] has been detected

in this plant, as have been glucosides,[216] chlorogenic acid,[217]

and o-protocatechuic acid,[218] the latter in acid hydrolysates.

Dambonitol,[46] robinoside[49] and ursolic acid[53, 219, 220]

are the only non-alkaloid constituents that have actually been

isolated from V. major.

E.   Alkaloids Isolated from Vinca major L.

The earliest investigations on the alkaloids of V. major

were by Janot and co-workers[221] in 1954.   This group suc-

ceeded in isolating reserpinine (11-methoxy-δ-yohimbine)

(L IX),   and a second alkaloid having m.p. 316°, which was

thought to be serpinine.   This report was followed in 1955, by

successive papers on the isolation of vincamajoridine,[222]

vincamajoreine,[223] and vincamajine[224] from this plant.

Vincamajoridine was subsequently shown to be identical with

the previously known alkaloid akuammine (L),[225] thus the latter

name should be preferred.

In 1960, Farnsworth et al.[75] reported on the isolation of

perivincine from V. major, which had previously only been

isolated from V. minor.[74] Trojákek et al.[76, 77] showed that

perivincine was a mixture of vincamine and vincine.[76]

Trojákek and Hodkova,[220] in 1962, confirmed the presence

of reserpinine and vincamajine in V. major, and also isolated

vincamedine, which had previously only been found in V.

difformis.[139, 226] The structures of vincamedine (XXXVIII)

and vincamajine (XXXVII) were subsequently deduced.[227, 228]

Russian workers[229, 230] next isolated a new base, majdine,

from V. major, and also confirmed the presence of reserpinine

and akuammine. The structure of majdine (LII)[231] was later

determined to be an oxindole base stereoisomeric with cara-

panaubine (LIII). The French group,[232] in 1965, isolated a new

base from V. major roots, and determined its structure as

10-methoxyvellosimine (XLII). In addition, vincamajoreine

(XXXIX) was shown to be a stereoisomer of 10-methoxytetra-

phyllicine,[233] and vincamine (I) and vincine (III) were isolated

for the first time from the aerial parts of V. major.[233] An

alkaloid designated as alkaloid V was shown to be a dimethoxy-

oxindole base, but its complete structure is still unreported.[233]

Kaul and Trojánek,[234, 235] working with the aerial parts

of V. major, isolated the new alkaloids majovine, majorine,

majoridine and vincanovine, and confirmed the presence of

majdine (majoroxine)(LII),[235] reserpinine, vincamajine and

vincamedine. The structure of majoridine (XL)[236] was subse-

quently shown to be 10-methoxy-O-acetyltetraphyllicine, and

its absolute configuration at C_{15} was determined and shown to

be consistent with that known for other indole and dihydroindole

alkaloids.

Reserpine (LI) has been identified in <u>V. major</u> by Borkow-

ski <u>et al.</u>[237] on the basis of thin-layer chromatography and UV

spectroscopy of amorphous fractions, but this identification

must be considered as equivocal. Farnsworth <u>et al.</u>,[210] on

the basis of paper and thin-layer chromatographic studies of

several alkaloid fractions from <u>V. major</u>, together with com-

parisons of R_F values of known compounds, stated that there

are at least 37 bases in this plant. They tentatively identified

four of these as sarpagine, tetrahydroalstonine, serpentine

and vincamine. However, of these four, only vincamine has

been isolated to date.

The physical data for all alkaloids isolated from <u>V. major</u>

are presented in Table 3, and their structures, where known

are shown in Charts 1-16.

F. <u>Distribution of Vinca major Alkaloids</u>

A total of 14 alkaloids have been isolated in crystalline

form from <u>V. major</u>, and the structures for all except four

(majorine, majovine, vincanovine, alkaloid V) have been

determined.

(+)-Vincamine (I) has been isolated from <u>V. minor</u>,[62] <u>V.</u>

<u>major</u>,[233] <u>V. difformis</u>,[139] <u>V. erecta</u>,[140-142] and racemic

vincamine as well as (+) vincamine have been isolated from

Tabernaemontana rigida[135]; vincine (III) from V. major[233] and

V. minor[77, 86, 87]; vincamajine (XXXVII) from V. major,[220,]

[224, 234] V. difformis,[226] Rauvolfia mannii,[238] and Tonduzia

longifolia[239]; 14-epivincamine from V. minor[70, 99, 101] and

Tabernaemontana rigida[135]; vincamedine (XXXVIII) from V.

major[220, 234] and V. difformis[226]; akuammine (vincamajoridine)

(L) from V. major,[222, 229, 230, 233, 234] V. erecta,[140]

Picralima klaineana (P. nitida),[241] and Catharanthus roseus

(Vinca rosea)[242, 243]; majdine (LII) from V. major[229, 230]

and V. herbacea[231, 244]; and reserpinine (LIX) from V. major,

[221, 229, 234] V. pubescens,[206, 230] V. erecta,[245] V. herbacea,

[244] and from at least 13 Rauvolfia species.[145, 154, 158, 172,]

[246, 249-252]

Majoridine, majorine, majovine, vincanovine, vinca-

majoreine, 10-methoxyvellosimine, and vincadiffine are

restricted in their occurrence to V. major.

Thus, the 14 alkaloids of V. major are distributed through-

out at least 24 species of six genera in the Apocynaceae, Vinca,

Tonduzia, Picralima, Catharanthus, Rauvolfia and Tabernae-

montana.

TABLE 3

Physical Data for Alkaloids Isolated from Vinca major L.

Alkaloid	Structure	Formula	m.p. (°C)	$[\alpha]_D$	λ_{max} (mμ)
10-Methoxyvellosimine[232]	XLII	$C_{20}H_{22}N_2O_2$	226	+71[d]	228, 280, 293
Vincamajoreine[223, 233]	XXXIX	$C_{21}H_{26}N_2O_2$	230 (block)	–	246, 310
			246-247 (cap.)	–	–
			271 (block)	–	–
Vincamine[233]	I	$C_{21}H_{26}N_2O_3$	274 (block)	+42[e]	225, 278
Vincamajine[220, 224, 233, 234]	XXXVII	$C_{22}H_{26}N_2O_3$	226-227	-22[d]	249, 292
			224	-55[c]	248, 291
			226.5-227	-54[c]	–
			225	-55[c]	248, 291
Akuammine[222, 225, 229, 234]	L	$C_{22}H_{26}N_2O_4$	265-270(d)	-64[d]	245, 312
			259 (cap.)	-67[e]	243, 320
			250-251	-103.2[e]	244, 312
			252(d) (cap.)	-104	245, 312

Table 3 (cont.)

Compound	Structure	Formula	m.p.	$[\alpha]$	UV maxima
Reserpinine[220, 221, 229, 232-234]	LIX	$C_{22}H_{26}N_2O_4$	241-242(d)	-121[d]	219, 297
			242	-109[e]	229, 299, 250(infl.)
			240-241	-128[d]	-
			240-241	-134[f]	-
			241-242(d)	-102[e]	-
Perivincine[75], g	-		199.5-200	-61[e]	-
Vincine[233]	III	$C_{22}H_{28}N_2O_4$	210(Kofler)	+39[e]	230, 270
Vincamedine[234, 236]	XXXVIII	$C_{22}H_{28}N_2O_4$	185-187	-75[d]	248, 292
Majoridine[234, 236]	XL	$C_{23}H_{28}N_2O_3$	222-223	-26.6[d]	247, 310
Majdine[227, 228] (majoroxine)	LII	$C_{23}H_{28}N_2O_6$	192-194	-141[d]	224, 248(infl.)
			186-188(d)	-137[b]	-
Alkaloid V[233]	-	$C_{23}H_{28}N_2O_6$	Amorph.	-	224, 280, 255(infl.)
Majorine[234, 235]	-	$C_{24}H_{26}N_2O_3$	247-249(d)	-265[d]	223, 289
Majovine[234, 235]	-	-	227-229	+133[d]	221, 272
Vincanovine[234, 235]	-	-	330(d)	-20[c]	232, 327
Serpinine (?)[221]	-	-	316	-	-

[b]Methanol, [c]Ethanol, [d]Chloroform, [e]Pyridine, [f]Acetone, [g]Claimed to be a mixture of vincamine and vincine[76, 77].

[a]For structures see Figures 1, 9-11, 13 and 14.

IV. THE PHYTOCHEMISTRY OF VINCA PUBESCENS URV.

A. Description and Occurrence

Vinca pubescens Urv. is considered by Pichon[1] to be synonymous with V. major var. major Pich. However, recently this has been disputed by Pobedimova,[197] who considers
it a distinct species on the basis of morphological characteristics and distribution. Other names that have appeared in the
literature for this species are Vinca major Ldb., V. major L.
V. pubescens, and V. major L. var. major Pich.

This species is recorded by Pobedimova as growing in
shady forests near the coastline in the Caucasus, and spreading
to the Balkans and Asia minor. The vegetative stems are
creeping and not rooted, whereas the flowering branches are
usually erect. It has ovoid leaves, 5 to 6 cm. long and 2 to 3.5
cm. wide, which are rounded at the base or slightly cordate,
and which have erect hairs. The flower is azure, single,
axillary and 20 to 30 mm. in diameter.

B. Folklore and Biological Properties

No references could be found that associate V. pubescens
with medicinal folklore.

Orechoff et al.[206] reported hypotensive effects for alkaloid fractions from this plant in laboratory animals and one of the

alkaloids isolated in their studies, vinine (carapanaubine), was

shown to elicit hypotensive activity. Pubescine (reserpinine)

elicits only a transient hypotensive effect in laboratory

animals.[212]

C. Non-Alkaloid Constituents of Vinca pubescens Urv.

The only non-alkaloid constituent reported from V. pubes-

cens is ursolic acid, which was obtained in a yield of 0.57 per

cent.[230]

D. Alkaloids Isolated from Vinca pubescens Urv.

Orechoff et al.,[206] in 1934, isolated three alkaloids from

this plant utilizing a type of gradient pH fractionation of the

crude total bases. In this manner, they isolated pubescine,

vinine and a third alkaloid having m.p. 194-195°.

The identity of these bases remained obscure until Abdura-

khimova et al.[230] re-investigated this species, following the

procedures described by Orechoff,[206] and isolated pubescine

and vinine, which had higher melting points than the same bases

of Orechoff. Vinine was shown by these workers to be identical

with the oxindole base carapanaubine (LIII), and pubescine with

reserpinine (LIX). No direct comparison of Abdurakhimova's

bases, however, could be made with those of Orechoff, since the latter were no longer available.

Physical data for all V. pubescens alkaloids are presented in Table 4 and the structures for reserpinine and carpanaubine are shown in Charts 12 and 13.

E.   Distribution of Vinca pubescens Alkaloids

Carapanaubine (vinine) has not been reported from any Vinca species except V. pubescens, a fact that lends support to the contention of Pobedimova[197] that this species is separate and distinct from V. major L. This base has also been report- ed from Aspidosperma carapanauba,[254] Aspidosperma rigida,[255] and Rauvolfia vomitoria.[256]

Reserpinine, on the other hand, occurs not only in V. pubescens,[206, 230] but in V. major,[221, 229, 232-234] V. erecta,[245] and V. herbacea,[244] in addition to at least 13 Rauvolfia species.[145, 154, 158, 172, 246, 249-252]

Thus, the two alkaloids of know constitution from V. pubescens occur in at least 18 species of only two genera in the Apocynaceae, i.e. Vinca and Rauvolfia.

Only more detailed phytochemical studies on authentic V. pubescens Urv. will provide the information necessary to tell whether or not it is truly distinct from V. major L., at least on a chemical basis.

TABLE 4

Physical Data for Alkaloids Isolated from Vinca pubescens Urv.

Alkaloid	Structure[a]	Formula	m.p.(°C)	$[\alpha]_D$	λ_{max}(mμ)
Carapanaubine [206, 230]	LIII	$C_{19}H_{26}N_2O_4$	211.5-213 216-217	-70.12	-
Reserpinine [206, 230]	LIX	$C_{20}H_{26}N_2O_4$	227-228	-134.2	-
Alkaloid [206]	-	-	194-195	-	-

[a]For structures see Figures 12 and 13.

V.   THE PHYTOCHEMISTRY OF VINCA DIFFORMIS POURR.

A. Description and Occurrence

Vinca difformis Pourr. is indigenous to the Azores, Spain,
Portugal, France, Italy, Corsica, Sardinia, Balearic Islands,
Morocco, and Algeria. Botanical synonyms for this species
include Vinca media Hoffmgg. et Link, Vinca acutiflora Bertol.,
Vinca intermedia Tausch., Pervinca media (Hoffmgg. et Link)
Caruel, Vinca obtusiflora Pau and Vinca major L. var. glabra
F. Schultz.[1-3]

As with V. major L., V. difformis Pourr, exhibits a dimorphic growth, with sterile creeping, and an erect flowering
branch. The leaf is soft and obovate, and the leaves and sepals
are devoid of marginal hairs.[1-3]

Pichon considers V. difformis Pourr, as a simple variety
of Vinca major L., and has classified it as Vinca major L.
var. difformis (Pourr.) Pich.[1]

B. Folklore and Biological Properties

No folkloric applications have been found in the literature
regarding this species, and crude extracts have not been subjected to any type of biological evaluation. However, the plant
does contain vincamine, and its pharmacology[28-42] and clinical

applications[20-27] have been discussed previously. Limited

pharmacologic data are available on akuammidine,[257] and

sarpagine is a non-sedative sympatholytic base which gives

only a transient hypotensive effect in animals.[212, 258, 259]

C. <u>Non-alkaloid Constituents of Vinca difformis Pourr.</u>

This plant remains uninvestigated as regards non-alkaloid

entities. All phytochemical studies have been directed toward

the isolation of alkaloids.

D. <u>Alkaloids Isolated from Vinca difformis Pourr.</u>

All investigations on <u>V</u>. <u>difformis</u> alkaloids have been

carried out by Janot, LeMen and co-workers. The major

groups of alkaloids present in this plant are of the ajmaline-

type, represented by vincamajine (XXXVII),[261, 262] vinca-

medine (XXXVIII),[129, 226, 261, 262] as well as the sarpagine

type, represent by vellosimine (XLIII),[261] sarpagine (XLIV),[139]

and akuammidine (XLV).[260-262] Also isolated has been the

eburnamine base vincamine (I),[139] the aspidosperma racemate

(±)-vincadifformine (XIX$_{dl}$),[138, 260, 261, 265] and two 2-acyl-

indole bases, vincadiffine (LXIV),[261, 266] and an unidentified

base having m.p. 236-238°.[261]

Physical data for these alkaloids are presented in Table 5, and the structures, where known, are shown in Charts 1-16.

This plant has not yielded representatives of the quebrach-amine, aspidospermatine, oxindole or yohimbine ring E heter-ocycle-types of alkaloids, which are quite characteristics for other Vinca species. In this respect, it is quite clear that there is a definite chemical difference between V. major L. and V. difformis Pourr., which appears to justify their classi-fication as distinct species, rather than be considered as sim-ple varieties of V. major L.

E. Distribution of Vinca difformis Alkaloids

Sarpagine has been shown to be present in V. difformis[139] and in at least 10 different Rauvolfia species,[154, 161, 166, 178, 246-248, 267-271] but in no other plants. Vincamine has been isolated from V. difformis,[139] V. minor,[62] V. major,[233] V. erecta,[140-142] and from Tabernaemontana rigida[135]; vincama-jine occurs in V. difformis,[261, 262] V. major,[220, 224, 233, 234] Rauvolfia mannii,[238] and in Tonduzia longifolia[239]; akuammidine (rhazine, ervamidine) is present in V. difformis,[260, 261] V. erecta,[272] Rhazya stricta,[131, 133] Aspidosperma quebracho-blanco,[273] Melodinus australis,[125] Vallesia dichotoma[134] and

TABLE 5

Physical Data for Alkaloids Isolated from <u>Vinca difformis</u> Pourr.

Alkaloid	Structure	Formula	m.p.(°C)	$[\alpha]_D$	$\lambda_{max}(m\mu)$
Vellosimine[261]	XLIII	$C_{19}H_{20}N_2O$	304-306	+56[b]	227, 280, 289
Sarpagine[139]	XLIV	$C_{19}H_{22}N_2O_2$	>360	+55[e]	225, 280
Akuammidine[260, 262]	XLV	$C_{21}H_{24}N_2O_3$	241	+24[b]	227, 280
(±)-Vincadifformine[260, 265]	XIX_{dl}	$C_{21}H_{26}N_2O_2$	125	±0[b]	225, 300, 328
Vincamine[139]	I	$C_{21}H_{26}N_2O_3$	222(cap.)	+43[e]	225, 280
Vincamajine[261, 262]	XXXVII	$C_{22}H_{26}N_2O_3$	-	-	-
Vincadiffine[261, 266]	LXIV	$C_{22}H_{28}N_2O_4$	230 / 193-197	-121[d] / -115[c]	242, 320
Vincamedine[139, 226, 262]	XXXVIII	$C_{24}H_{28}N_2O_4$	185	-66[d]	250, 295
Alkaloid[261]	-	-	236-238	-	242, 310

aFor structures see Figures 1-16; bMethanol; cEthanol; dChloroform; ePyridine.

in Gonioma kamassi[120]; (+)-vincadifformine in V. difformis,[260,] [261, 265] V. minor,[95] Tabernaemontana riedelii,[135] Vallesia dichotoma,[134] and Rhazya stricta[123]; vincamedine in V. difformis[139, 226, 261] and in V. major[234] and vellosimine has been isolated from V. difformis,[261] Geissosperum vellosii[274] and Rauvolfia verticillata.[275]

Eight alkaloids of known constitution have been isolated from Vinca difformis Pourr. Six of these eight alkaloids are also present in other apocynaceous plants. Thus, the alkaloids of V. difformis are distributed in about 22 species of 9 genera in the Apocynaceae (Vinca, Rauvolfia, Melodinus, Tonduzia, Gonioma, Tabernaemontana, Rhazya, Aspidosperma, Geissosperum).

VI. THE PHYTOCHEMISTRY OF VINCA HERBACEA WALDST. ET KIT.

A. Description and Occurrence

Vinca herbacea Waldst. et Kit. is indigenous to southern Europe (lower Austria), through the middle East and southern Russia, as far South as Tukestan. It exhibits a low growth, seldom erect, with blue-violet flowers and soft leaves with un-even, rough margins, which are rarely ciliate, and are from

2 to 5 cm long. Botanical synonyms for this species are <u>Vinca</u>

<u>pumila</u> Clarke, <u>Vinca mixta</u> (Velen.) Velen. and <u>Vinca erecta</u>

Regel. et Schmalh. var. <u>bucharica</u> B. Fedtsch. Pichon con-

siders this plant to be <u>Vinca herbacea</u> Waldst. et Kit. var.

<u>herbacea</u> Pich.[1-3]

B. Folklore and Biological Properties

No folkloric uses could be found in the available literature

for <u>V. herbacea</u>. The alkaloids of this species have been

shown to be antimicrobial,[276] and several pharmacologic

studies have been reported for the crude bases which show

them to have hypotensive and muscle relaxant properties.[277-280] Lochnerinine is a cytotoxic alkaloid.[281]

C. Pharmacognostic Studies on Vinca herbacea Waldst. et Kit.

The pertinent, identifying, microscopic features of <u>V. herbacea</u> have been reported in studies by Blažek and Stary[44]

and by Bocharova.[282]

D. Non-alkaloid Constituents of Vinca herbacea Waldst. et Kit.

All phytochemical studies on <u>V. herbacea</u> have been con-

fined to the separation and identification of alkaloids, except

for one study in which ursolic acid was isolated from this

plant.[283]

E. Alkaloids Isolated from Vinca herbacea Waldst. et Kit.

All of the definitive alkaloid investigations on this plant

have been carried out by Ognyanov and co-workers in Bulgaria.

These studies were initiated in 1961. The majority of alkaloids

from V. herbacea can be placed in three major groups, i.e.

the yohimbinoid ring E heterocycles, including reserpinine

(LIX),[244] herbaine (LX),[283, 284] and herbaceine (LXI)[283, 285,
286]; the oxindole group, including majdine (LII),[231] isomajdine

(LIV),[231] and herbaline (LV)[286, 287]; and the aspidosperma

group, including lochnerinine (XXIV),[288] (-)-tabersonine (XXV),

[288] and 16-methoxy-(-)-tabersonine (XXVI).[288]

Norfluorocurarine (vincanine) (XXXIII)[289] is the only represent-

ative of the aspidospermatine group found in this plant, and

hervine (LXV)[290] represents the yohimbinoid seco ring E

group of alkaloids.

Two uncharacterized alkaloids named vincaherbine and

vincaherbinine were isolated from this plant in 1963 by

Zabolotnaya and Bukreeva.[291] These bases were subsequently

reinvestigated and named herbaine and herbaceine respectively.
[284]

The physical data for all V. herbacea alkaloids are pre-

sented in Table 6, and the structures, where known, are

shown in Charts 1-16.

F. <u>Distribution of Vinca herbacea Alkaloids</u>

It is interesting to note that no alkaloids of the eburnamine

or quebrachamine types are found in <u>V. herbacea</u>. Of the alka-

loids of the aspidosperma-type, lochnerinine is found in <u>V.</u>

<u>herbacea,</u>[288] <u>Catharanthus roseus (Vinca rosea),</u>[292, 293]

<u>Catharanthus lanceus</u>[294] and in <u>Catharanthus pusillus</u>[295]; (-)-

tabersonine from <u>V. herbacea,</u>[288] <u>Amsonia angustifolia,</u>[296]

<u>Conopharyngia durissima,</u>[297] <u>Amsonia tabernaemontana,</u>[298]

<u>Stemmadenia donnell-smithii,</u>[299] <u>Stemmadenia obovata,</u>[299]

<u>Stemmadenia tomentosa var. palmeri,</u>[299] and <u>Tabernaemontana</u>

<u>alba</u>[299]; and 16-methoxy-(-)-tabersonine only from <u>V. herbacea.</u>

[288] <u>Norfluorocurarine,</u> an aspidospermatine-type base, is

identical with the alkaloid vincanine from <u>V. erecta,</u>[300] which

has also been isolated from <u>V. herbacea</u>[289] and <u>Diplorrhynchus</u>

<u>condylocarpon</u> ssp. <u>mossambicensis.</u>[301] Of the oxindole bases,

majdine has been isolated only from <u>V. herbacea</u>[231] and <u>V.</u>

<u>major,</u>[229, 230] and herbaline[287] and isomajdine[231] are found

only in <u>V. herbacea.</u> The yohimbinoid heterocyclic ring E

bases herbaine[283, 284] and herbaceine[283-286] are restricted in

their occurrence to <u>V. herbacea,</u> whereas reserpinine has been

isolated from <u>V. herbacea,</u>[244] <u>V. major,</u>[221, 229, 230, 232-234]

<u>V. pubescens,</u>[206, 230] <u>V. erecta,</u>[245] and from at least 13

TABLE 6

Physical Data for Alkaloids Isolated from Vinca herbacea Waldst. et Kit.

Alkaloid	Structure[a]	Formula	m.p.(°C)	$[\alpha]_D$	λ_{max}(mμ)
Norflurocurarine[289] (vincanine)	XXXIII	$C_{19}H_{22}N_2O$	182-186(d)	-1248[d]	-
(-)-Tabersonine[288]	XXV	$C_{21}H_{24}N_2O_2$	192(d)(HCl)	-366[c]	225, 300, 327
Alkaloid A-1[244]	-	$C_{22}H_{24}N_2O_3$	228-231	-	-
16-Methoxy-(-)-tabersonine[288]	XXVI	$C_{22}H_{26}N_2O_3$	184-186(d)(HCl) -310[d]		245, 321
Reserpinine[244] (pubescine)	LIX	$C_{22}H_{26}N_2O_4$	242-244	-	-
Lochnerinine	XXIV	$C_{22}H_{26}N_2O_4$	Amorph.	-	247, 327
Hervine[290]	LXV	$C_{22}H_{28}N_2O_4$	173-175	-93[c]	228, 270, 298
Herbaine[283,284] (vincaherbine)	LX	$C_{22}H_{28}N_2O_4$	126-128(d) 129-130	-218[e] -253	228, 274, 297 -

Name		Formula	mp (°C)	$[\alpha]_D$	UV λ_{max}
Majdine [231] (Alkaloid A-5)	LII	$C_{23}H_{28}N_2O_6$	192-194 190-192	- -108^e	224, 248
Isomajdine [231] (Alkaloid A-4)	LIV	$C_{23}H_{28}N_2O_6$	208-210	-111^e	-
Herbaceine [283-286] (vincaherbinine)	LXI	$C_{23}H_{30}N_2O_5$	144(d) 139-140 144-145	-219^d -238^d -219	226, 300, 280(sh) 228, 270, 298 226, 252
Herbaline	LV	$C_{23}H_{30}N_2O_6$	276-278(d)	-147^e	215, 273, 305
Alkaloid A-3	-	-	232-235	-94^e	250, 298

[a] For structures see Figures 1-16; [b] Methanol; [c] Ethanol; [d] Chloroform; [e] Pyridine.

Rauvolfia species.[145, 154, 158, 172, 246, 249-252] Hervine is the

only yohimbinoid seco ring E type base found in any Vinca

species, and it is restricted to V. herbacea.[290]

Eleven alkaloids of known structure have been isolated

from V. herbacea. These 11 bases are distributed in about 28

species of eight genera in the Apocynaceae, i.e. Vinca, Catha-

ranthus, Tabernaemontana, Stemmadenia, Rauvolfia, Cono-

pharyngia, Diplorrhynchus, Amsonia.

VII. THE PHYTOCHEMISTRY OF VINCA ERECTA REGEL ET SCHMALH.

A. Description and Occurrence

Vinca erecta Regel et Schmalh. is a low growing, seldom

erect plant, with blue-violet flowers and oblong, soft leaves

without indented margins. It is indigenous to Turkestan,

Syria, Lebanon, Cilicia and Mesopotamia. Botanical synonyms

for this species include Vinca libanotica Zucc., Vinca bottae

Juab et Spach and Vinca herbacea Waldst. et Kit. var. glaber-

rima A.DC. Pichon considers this plant as Vinca herbacea

Waldst. et Kit. var. libanotica (Zucc.) Pich.[1-3]

B. Folklore and Biological Properties

No folkloric uses could be found in the available literature

for Vinca erecta.

The general pharmacology for normacusine B (tombozine), ervinine,[302] [303-305] norfluorocurarine (vincanine),[306-312] vincanidine,[306] ervamine,[313, 314] akuammidine (ervamidine), vinervine[315] and vineridine[316] [317] as well as for the total alkaloids of Vinca erecta,[314] has been described, and the structure-activity relationships of four derivatives of vincanine have been reported.[318] Typical actions are hypotensive and tranquilizing effects, increased amplitude and frequency of heart contractions, increased coronary heart flow, contraction of uterine musculature and hemostatic effect.

Vincanine (HCl), vincanidine (HCl), ervamine (HBr), ervamine (HI), and reserpinine were evaluated for antimicrobial properties by Abidov et al.[319] Vincanine (norfluorocurarine) and vincanidine inhibited the growth of hemolytic streptococci, as well as Staphylococcus aureus and S. albus, and dysentery Gram negative rods, at dilutions of 1-2000 to 1-8000. Ervamine inhibited the growth of Pseudomonas aeruginosa, S. aureus and S. albus at a dilution of 1-1000.

C. Non-alkaloid Constituents of Vinca erecta Regel et Schmalh.

The only non-alkaloid isolated from V. erecta is ursolic acid.[320]

D. Alkaloids of Vinca erecta Regel et Schmalh.

The alkaloid isolation, identification and structure elucidation studies on V. erecta have been carried out by Yunusov, Yuldashev and co-workers.

Several different groups of alkaloids are present in V. erecta. In this respect, it is perhaps the most diverse member of the genus Vinca. A total of 23 alkaloids have been isolated to date from this plant. Five of these (vincanidine, norfluorocurarine, vincamine, vincarine, reserpinine) occur in the roots, but only reserpinine is restricted to the roots. The structures for 20 of the 23 alkaloids are known. Ercine,[321] ervinidine[322] and ervinidinine[322] are of unknown structure at present. Vincanine (norfluorocurarine)(XXXIII) is the major root alkaloid,[300, 323, 324] and kopsinine (XXXI) is the major alkaloid from the aerial parts,[140, 325] although vinerine (LVI) and vineridine (LVII) are also present in high concentrations in the aerial parts.[140, 326] Two yohimbinoid ring E heterocycles, ervine (LXII)[327] and reserpinine (LIX), [245] have been isolated from this plant, as have three oxindole bases, vinerine (LVI), [139, 140, 326] vineridine (LVII),[139, 140, 326] and ercinine (LVIII). [348] Four aspidospermatine alkaloids occur in V. erecta, namely, vinervine(XXXIV),[139, 140, 328-330] vinervinine (XXXV),[331, 332] vincanine (XXXIII),[142, 143] and vincanidine (XXXVI).[140, 141, 323, 325, 337, 330, 333, 334]

Vincanine (XXXIII) was shown to be identical with norfluoro-curarine.[141] Several sarpagine and sarpagine-like alkaloids have been isolated from this plant, including akuammidine (ervamidine) (XLV),[327] normacusine B (tombozine)(XLVI),[302] akuammine (vincamajoridine)(L),[140] ervincine (XLVIII),[335] and vincaridine (XLIX).[336]

Four aspidosperma-type bases have been isolated from V. erecta, namely, kopsinine (erectine)(XXXI),[140, 325, 337] kopsinilam (XXXII),[335] ervamine (XXVII),[338] and Ψ-kopsinine (XXVIII).[140, 325, 330, 331] The latter base is identical with dihydrovindolinine.[332]

Only one representative of the eburnamine group of alka-loids has been isolated from V. erecta, and that is vincamine (I).[140, 323] Vincarine (XLI),[339-341] a diastereoisomer of quebrachidine, is representative of ajmaline-type alkaloids of V. erecta. No quebrachamine bases have as yet been reported to occur in this plant. In fact, this type of alkaloid appears to be curiously restricted, within the genus Vinca, to V. minor.

A method for identifying most of the alkaloids of V. erecta, utilizing thin-layer chromatography in conjunction with the ceric ammonium sulfate chromogenic spray reagent, has been published.[342]

Finally, there is evidence reported that V. erecta is a

TABLE 7

Physical Data for Alkaloids Isolated from Vinca erecta Regel. et Schmalh.

Alkaloids	Structure[a]	Formula	m.p.(°C)	$[\alpha]_D$	λ_{max}(mμ)
Norfluorocurarine[141,245,323,324] (vincanine)	XXXIII	$C_{19}H_{20}N_2O$	210-211(HCl); 187.5-188(base)	-985.8; -992[b]	245, 298, 370; -
Normacusine B[302] (tombozine)	XLVI	$C_{19}H_{22}N_2O$	-	-	-
Vincanidine[245,323,325,329,330,333,334]	XXXVI	$C_{19}H_{20}N_2O_2$	250-280(d)	-848[b]	
Vinervine[140,328-330]	XXXIV	$C_{20}H_{24}N_2O_3$	154-155(d)	-505[b]	234, 290, 336
Vincaridine[336]	XLIX	$C_{20}H_{24}N_2O_3$	216-217	-58	234, 238
Ervinidinine[322]	-	$C_{21}H_{24}N_2O_3$	255-258(d)	-160.6[b]	-
Ervine[327]	LXII	$C_{21}H_{24}N_2O_3$	222-223	-57.3[b]	227, 282, 291
Akuammidine[327] (ervamidine)	XLV	$C_{21}H_{24}N_2O_3$	242-243	+26.9[b]	227, 281
Vincarine[339-341]	XLI	$C_{21}H_{24}N_2O_3$	263-264	+13.98[b]	242, 292

Ervincine[335]	XLVIII	$C_{21}H_{24}N_2O_3$	156-157	+93[d]	228, 276, 316
Kopsinilam[335]	XXXII	$C_{21}H_{24}N_2O_4$	248-249	-13.5[d]	246, 295
Vinervinine[331,332]	XXXV	$C_{21}H_{24}N_2O_4$	190-191	-564[d]	237, 292, 334
Ervamine[338]	XXVII	$C_{21}H_{26}N_2O_2$	198-200(H1)	-	-
Kopsinine[140,325] (erectine)	XXXI	$C_{21}H_{26}N_2O_2$	136-138	-30.4	210, 250, 295
Ψ-Kopsinine[140,325,332]	XXVIII	$C_{21}H_{26}N_2O_2$	-	-	226, 247, 297
Vincamine[140,245,323] (minorine)	I	$C_{21}H_{26}N_2O_3$	224-225	-	-
Ericinine[348]	LVIII	$C_{21}H_{26}N_2O_5$	206-207	+43.8[f]	220
Ervinidine[322]	-	$C_{22}H_{26}N_2O_4$	283-284(d)	-17.3	232, 302, 340
Reserpinine[245]	LIX	$C_{22}H_{26}N_2O_4$	236-237	-134	-
Akuammine[140] (vincamajoridine)	L	$C_{22}H_{26}N_2O_4$	-	-	-
Vinerine[140,240,326]	LVI	$C_{22}H_{26}N_2O_5$	202-203	+20.3[e]	220

Vineridine[140, 240, 326]	LVII	$C_{22}H_{26}N_2O_5$	179-180	+22.7[e]	220
Ercine[321]	-	$C_{25}H_{28}N_2O_6$	158-159	-121[b]	235, 290

[a]For the structures see Figures 1-16; [b]Methanol; [c]Ethanol; [d]Chloroform; [e]Pyridine; [f]Acetone.

plant in which the alkaloids differ quantitatively as well as qualitatively, depending on ecologic factors.[240]

The physical data for all V. erecta alkaloids are presented in Table 7, and structures for these alkaloids, where known, are shown in Charts 1-16.

E. Distribution of Vinca erecta Alkaloids

Of 23 V. erecta alkaloids, five occur in other Vinca species, whereas seven occur in genera other than Vinca in the Apocynaceae.

Norfluorocurarine(vincanine) is found in V. erecta,[142, 300, 323, 324] V. herbacea,[288] Geissospermum vellosii[274] and in Diplorrhynchus condylocarpon ssp. mossambicensis[301]; reserpinine in V. erecta,[245] V. herbacea,[244] as well as in at least 13 Rauvolfia species.[145, 154, 158, 172, 246, 249-252] Vincamine is restricted to V. erecta,[140-142] V. minor,[62] V. major,[233] V. difformis,[139] and Tabernaemontana riedelii.[135]

Akuammidine (rhazine, ervamidine) occurs in V. erecta,[272] in V. difformis,[260, 261, 262] in Rhazya stricta,[131, 133] Aspidosperma quebracho-blanco,[273] Melodinus australis,[125] Vallesia dichotoma,[134] and in Gonioma kamassi.[121]

Kopsinine is present in V. erecta,[140, 325] in Alstonia venenata,[194, 195, 343] and in Kopsia longiflora,[344] Pleiocarpa flavescens,[345] Pleiocarpa pycnantha var. tubacina,[346]

Pleiocarpa mutica,[117] and in Aspidosperma populifolium[253]

and Aspidosperma multiflorum.[253] Kopsinilam is present in

V. erecta,[335] Hunteria eburnea,[345] and Pleiocarpa mutica.[345]

Akuammine (vincamajoridine) has been isolated from V.

erecta,[140] V. major,[222, 229, 230, 233, 234] Picralima

klaineana (P. nitida),[241] and Catharanthus roseus (Vinca

rosea).[242, 243]

Normacusine B (tombozine, vellosiminol) occurs in V.

erecta,[302] Aspidosperma polyneuron,[128] Geissospermum

vellosii,[274] Diplorrhynchus condylocarpon ssp. mossambi-

censis,[301] and in Rauvolfia perakensis.[347]

The alkaloids of V. erecta are distributed in at least 36

species of 15 genera of the Apocynaceae, i.e. Vinca, Catha-

ranthus, Rauvolfia, Hunteria, Aspidosperma, Geissospermum,

Pleiocarpa, Diplorrhynchus, Kopsia, Rhazya, Tabernaemon-

tana, Picralima, Alstonia, Melodinus and Gonioma, thus

making them the most widespread in the Apocynaceae of any

Vinca species.

VIII. SUMMARY

Twenty-four of the 86 alkaloids from Vinca species occur

in other apocynaceous plants. Those plants in which the

alkaloids occur are distributed in 20 genera of four tribes

classifed in the sub-family Plumerioideae. The genera

include <u>Alstonia</u>, <u>Amsonia</u>, <u>Aspidosperma</u>, <u>Catharanthus</u>,

<u>Conopharyngia</u>, <u>Diplorrhynchus</u>, <u>Geissospermum</u>, <u>Gonioma</u>,

<u>Haplophyton</u>, <u>Hunteria</u>, <u>Kopsia</u>, <u>Melodinus</u>, <u>Picralima</u>, <u>Pleio-</u>

<u>carpa</u>, <u>Rauvolfia</u>, <u>Rhazya</u>, <u>Stemmadenia</u>, <u>Tabernaemontana</u>,

<u>Tonduzia</u> and <u>Vinca</u>. Exhaustive phytochemical studies on the

alkaloids of this group of plants, except for <u>Vinca pubescens</u>

Urv. have been performed, and some 64 of these are unique

to the genus.

Very little work has been done on the isolation of non-

alkaloid entities from <u>Vinca</u> species. No work of this nature

has been reported for <u>V. difformis</u>. Ursolic acid has been

isolated from all other <u>Vinca</u> species, and only scattered re-

ports list dambonitol, ornol, fructose, sucrose, robinoside,

rubber, 3-β-D-glucosyloxy-2-hydroxybenzoic acid, β-sitoster-

ol and <u>n</u>-triacontane from <u>V. minor</u>.

Pharmacologic data, and other biological effects, have

been reported for 13 of the <u>Vinca</u>-derived alkaloids, viz.

vincamine (minorine), carapanaubine (vinine), reserpinine

(pubescine), akuammidine (ervamidine), sarpagine, lochneri-

nine, <u>normacusine-B</u> (tombozine), ervinine, <u>norfluorocurarine</u>

(vincanine), vincanidine, ervamine, vinervine and vineridine.

Vincamine is the only alkaloid derived from Vinca species

that is used clinically.

ADDENDA

After this chapter was prepared, additional substances

were reported isolated from several of the Vinca species.

These are briefly noted below.

Vinca minor- Robinin, kampferol glycoside, quercetin

glycoside[349], vincasine[350], 16-methoxyvincadifformine[351]

16-methoxyminovincinine[352], and an isomer of (-)-minovinci-

nine[353] have been isolated from V. minor, and the structure

of vincatine was elucidated, showing it to be a new oxindole

[350]. This then brings to 40 the number of alkaloids isolated

from V. minor.

Vinca herbacea- Rutin[354, 355], a new α-methyleneindoline

alkaloid of unknown structure[356], and the new alkaloid her-

barinine[357], have been reported recently as having been iso-

lated from V. herbacea. Sixteen alkaloids have now been re-

ported from this species, and all but four have been assigned

structures.

Vinca erecta- The new alkaloids erysinine[358] ervinceine

[359], ervamicine[359], and vincanicine[360], were reported recent-

ly from this plant, and the structures for ervinidinine[361], ervsinine[358], ervinceine[359, 362], vincanicine[360], and ervamicine[358, 363] have been postulated. This brings to 27 the number of alkaloids reported from <u>V</u>. <u>erecta</u>. Only the alkaloids ervinidine[322] and ercine[321] of this group remain to have structures assigned.

REFERENCES

1. M. Pichon, Bull. Mus. Hist. Nat., <u>23</u>, 439 (1951).

2. N. G. Bisset, Ann. Bogor. <u>3</u>, 105 (1958).

3. R. Paris and H. Moyse, J. Agr. Trop. Botan. Appl., <u>4</u>, 481 (1957).

4. R. F. Carlson, Am. Nurseryman, <u>104</u>, 12 (1956).

5. J. Rabate, Bull. Soc. Chim. Biol. <u>15</u>, 1300 (1933).

6. R. T. Gunther, "Greek Herbal of Dioscorides", Hafner Publishing Co., New York, 1959, p. 404.

7. E. Szczeklik, J. Hano, B. Bogdanikow and J. Maj, Polski Tygod. Lekar., <u>12</u>, 121 (1957).

8. L. Aldaba and L. Oliveros-Belardo, Rev. Filip. Med. Farm., <u>29</u>, 259 (1938).

9. J. Hano and J. Maj, Acta Polon. Pharm., <u>14</u>, 171 (1957).

10. J. Hano and J. Maj, Polski Tygod. Lekar., <u>11</u>, 3 (1956).

11. G. G. Chatelier and E. S.rasky, Ann. Pharm. Franc., <u>14</u>, 677 (1956).

12. A. Quevauviller, J. Le Men and M.-M. Janot, Compt. Rend. Soc. Biol., <u>148</u>, 1791 (1954).

13. A. Quevauviller, J. Le Men and M.-M. Janot, Ann. Pharm. Franc., <u>12</u>, 799 (1954).

14. D. K. Zheliazkov, Suvremenna Med., 9, 16 (1958).

15. R. E. Heal, E. F. Rogers, R. T. Wallace and O. Starnes,
    Lloydia, 13, 89 (1950).

16. A. Frisbey, J. M. Roberts, J. C. Jennings, R. Y.
    Gottshall and E. H. Lucas, Quart. Bull., 35, 392 (1953).

17. W. H. Orgell, Lloydia, 26, 59 (1963).

18. G. F. McKenna, A. Taylor and B. S. Gibson, Texas
    Rept. Biol. Med., 18, 233 (1960).

19. B. J. Abbott, J. Leiter, J. L. Hartwell, M. E. Caldwell
    and S. A. Schepartz, Cancer Res., 26, 587 (1966).

20. F. Hargitai and J. Werkner, Orvosi Hetilap, 103, 312
    (1962).

21. J. Szabó and Z. Nagy, Arzneimittel-Forsch., 10, 811
    (1960).

22. P. Gömori, E. Gláz and Z. Szabó, Orv. Hetilap, 101,
    361 (1960).

23. I. Müller, Orv. Hetilap, 102, 359 (1961).

24. M. Földi and F. Obal, Therapia Hung., 13, 95 (1965).

25. A. Szabor, Therapia Hung. 13, 105 (1965).

26. B. E. Votchal and G. E. Tchapidze, Therapia Hung., 13,
    91 (1965).

27. A. Ravina, Presse Med., 74, 525 (1966).

28. R. -Hamet, Compt. Rend. Soc. Biol., 148, 1082 (1954).

29. Z. Szporny and K. Szász, Acta Physiol. Acad. Sci. Hung.,
    14, 46 (1958).

30. I. Krejčí, Cesk. Fysiol., 8, 452 (1959).

31. L. Szporny and K. Szász, Arch. Exptl. Pathol. Pharma-
    kol., 236, 296 (1959).

32. A. Káldor and Z. Szabó, Experientia, 16, 547 (1960).

33. F. Kaczmarek, J. Lutomski and T. Wrociński, Biul. Inst. Roslin Leczniczych, 8, 12 (1962).

34. O. Linét, I. Krejčí and M. Háva, Acta Biol. Med. Germ., 9, 158 (1962).

35. J. Machova and L. Macho, Biologia, 17, 456 (1962).

36. J. Molnár and L. Szporny, Acta Physiol. Acad. Sci. Hung. 21, 169 (1962).

37. F. Solti, Orv. Hetilap, 103, 202 (1962).

38. L. Szporny and P. Görög, Arch. Intern. Pharmacodyn., 138, 451 (1962).

39. J. Machova and F. V. Selecký, Bratislav. Lekarske Listy, 43, 449 (1963).

40. J. Machova and J. Mokrý, Arch. Intern. Pharmacodyn., 150, 516 (1964).

41. L. Szporny and P. Görög, Conf. Hung. Therap. Invest. Pharmacol., 2, Budapest, 1962, 237 (publ. 1964).

42. J. Machová, Arch. Intern. Pharmacodyn., 165, 459 (1967).

43. Z. Blažek and F. Starý, Cesk. Farm., 13, 153 (1964).

44. Z. Blažek and F. Starý, Cesk. Farm., 13, 165 (1964).

45. Z. Blažek and F. Starý, Cesk. Farm., 13, 315 (1964).

46. V. Plouvier, Compt. Rend., 251, 131 (1960).

47. J. Le Men and Y. Hammouda, Ann. Pharm. Franc., 14, 344 (1956).

48. F. E. King, J. H. Gilks and M. W. Partridge, J. Chem. Soc., 1955, 4206.

49. J. Rabate, Bull. Soc. Chim. Biol., 15, 130 (1933).

50. I. Hájková, V. Homola and L. Navratil, Pharmazie, 9, 537 (1959).

51. J. H. Mitchell, M. A. Rice and D. B. Boderick, Science, 95, 624 (1942).

52. J. Le Men and H. Pourrat, Ann. Pharm. Franc. 10, 349 (1952).

53. H. Pourrat and J. Le Men, Ann. Pharm. Franc. 11, 190 (1953).

54. J. Trojánek, J. Hoffmannová, O. Štrouf and Z. Čekan, Collection Czech. Chem. Commun., 24, 526 (1959).

55. R. K. Ibrahim and G. H. N. Towers, Arch. Biochem. Biophys., 87, 125 (1960).

56. R. Paris and H. Moyse-Mignon, Ann. Pharm. Franc., 14, 464 (1956).

57. G. H. Beale, J. R. Price and V. C. Sturgess, Bull. Roy. Soc., 130B, 113 (1941).

58. F. Rutishauser, Compt. Rend., 195, 75 (1932).

59. F. Rutishauser, Bull. Sci. Pharmacol., 39, 475 (1932).

60. J. Vintilesco and N. I. Ioanid, Bull. Soc. Chim. Romania, 14, 12 (1932).

61. E. S. Zabolotnaya, Tr. Vses. Nauchn. Issled. Inst. Lekarstv. i Aromat. Rast., 1950 (10), 29.

62. E. Schlittler and A. Furlenmeier, Helv. Chim. Acta, 36, 2017 (1953).

63. Z. Čekan, J. Trojánek and E. S. Zabolotnaya, Tetrahedron Letters, 1959, 11.

64. J. Trojánek, C. Štrouf, J. Holubek and Z. Čekan, Tetrahedron Letters, 1961, 702.

65. J. Mokrý, I. Kompiš and P. Šefčovič, Tetrahedron Letters, 1962, 433.

66. M. Plat, D. D. Mann, J. Le Men, M. -M. Janot, H. Budzikiewicz, J. M. Wilson, L. J. Durham and C. Djerassi, Bull. Soc. Chim. France, 1962, 1082.

67. C. Clauder, K. Gesztes and K. Szász, Tetrahedron Letters, 1962, 1147.

68. J. Mokrý, I. Kompiš, J. Suchý, P. Šefčovič and Z. Čekan, Collection Czech. Chem. Commun., 29, 433 (1963).

69. J. Trojánek, O. Štrouf, J. Holubek and Z. Čekan, Collection Czech. Chem. Commun., 29, 433 (1964).

70. J. Trojánek, Z. Koblicová and K. Bláha, Chem. Ind. (London), 1965, 1261.

71. J. Trojánek, Z. Koblicová and K. Bláha, Abhand. Deut. Wissen. zu Berlin, 3 Intern. Symp. Biochem. Physiol. Alkaloids, June, 1965 (publ. 1966).

72. M. E. Kuehne, J. Am. Chem. Soc., 86, 2946 (1964).

73. M. E. Kuehne, Lloydia, 27, 435 (1964).

74. Schiendlin, S. and N. Rubin, J. Am. Pharm. Assoc., Sci. Ed., 44, 330 (1955).

75. N. R. Farnsworth, F. J. Draus, R. W. Sager and J. A. Bianculli, J. Am. Pharm. Assoc., Sci. Ed., 49, 589 589 (1960).

76. We supplied Dr. Trojánek with a sample of perivincine which was isolated from V. major. In our hands this sample produced only one spot by paper chromatography. However, on re-examination of this alkaloid by us in a subsequent study, it was shown to be a mixture of vinca-mine and vincine by means of thin-layer chromatography.

77. J. Trojánek, K. Kavková, O. Štrouf and Z. Čekan, Collection Czech. Chem. Commun., 26, 867 (1961).

78. K. Szász, L. Szporny, E. Bittner, I. Gyenes, E. Hável and I. Magó, Kem. Folyoirat, 64, 296 (1958).

79. P. M. Lyapunova and Yu. G. Birosyak, Farm. Zh. (Kiev), 16, 48 (1961).

80. P. M. Lyapunova and Yu. G. Birosyak, Farm. Zh. (Kiev), 16, 42 (1961).

81. J. Trojánek, O. Štrouf, K. Kavková and Z. Čekan, Collection Czech. Chem. Commun., 25, 2045 (1960).

82. Z. Čekan, J. Trojánek, O. Štrouf and K. Kavková, Pharm. Acta Helv., 35, 96 (1960).

83. J. Trojánek, O. Štrouf, K. Kavková and Z. Čekan, Chem. Ind. (London), 1961, 790.

84. J. Trojánek, O. Štrouf, K. Kavková and Z. Čekan, Collection Czech. Chem. Commun., 27, 2801 (1962).

85. D. Zachystalová, O. Štrouf and J. Trojánek, Chem. Ind. (London), 1963, 610.

86. O. Štrouf and J. Trojánek, Chem. Ind. (London), 1962, 2037.

87. O. Štrouf and J. Trojánek, Collection Czech. Chem. Commun., 29, 447 (1964).

88. J. Holubek, O. Štrouf, J. Trojánek, A. K. Bose and E. R. Malinowski, Tetrahedron Letters, 1963, 897.

89. J. Trojánek, O. Štrouf, K. Bláha, J. Dolejš and V. Hanuš, Collection Czech. Chem. Commun., 29, 1904 (1964).

90. J. Mokrý, I. Kompiš, O. Bauerová, J. Tomko and Š. Bauer, Experientia, 17, 354 (1961).

91. J. Mokry, I. Kompis, P. Sefcovic, and S. Bauer, Collection Czech. Chem. Commun., 28, 1309 (1963).

92. J. Mokry, L. Dubravkova, and P. Sefcovic, Experientia, 18, 564 (1962).

93. J. Mokry, I. Kompis, L. Dubravkova, and P. Sefcovic, Tetra-
    hedron Letters, 1962, 1185.

94. J. Mokry, I. Kompis, M. Shamma, and R. J. Shine, Chem.
    Ind. (London), 1964, 1988.

95. J. Mokry, I. Kompis, L. Dubravkova, and P. Sefcovic,
    Experientia, 19, 311 (1963).

96. J. Mokry and I. Kompis, Naturwissenschaften, 50, 93 (1963).

97. I. Kompis and J. Mokry, personal communication.

98. J. Mokry and I. Kompis, Chem. Zvesti, 17, 852 (1963).

99. J. Mokry and J. Kompis, Tetrahedron Letters, 1963 (1970).

100. J. Mokry and I. Kompis, Lloydia, 27, 428 (1964).

101. J. Mokry, I. Kompis, and G. Spiteller, Collection Czech.
     Chem. Commun., 32, 2523 (1967).

102. I. Kompis and E. Grossmann, personal communication.

103. M. Plat, E. Fellion, J. Le Men, and M.-M. Janot, Ann.
     Pharm. Franc, 20, 899 (1962).

104. M. Plat, J. Le Men, M.-M. Janot, H. Budzikiewicz, J. M.
     Wilson, L. J. Durham, and C. Djerassi, Bull. Soc. Chem.
     France, 1962, 2237.

105. J. Le Men, J. Garnier, and M. Plat, Ann. Univ. Assoc.
     Reg. Etude Rech. Sci., Reims Sci. Med. Pharm., 3, 97
     1965).

106. W. Döpke and H. Meisel, Pharmazie, 21, 444 (1966).

107. W. Döpke, H. Meisel and G. Spiteller, Pharmazie,
     23, 99 (1968).

108. W. Döpke, H. Meisel, E. Gründeman and G. Spiteller,
     Tetrahedron Letters, 1968, 1805.

109. M. Debska and F. Kaczmarek, Planta Med., 7, 241
     (1959).

110. K. Szász, T. Kováts, M. E. Karácsony, Cs. Lörincz and J. Bayer, Planta Med., 7, 234 (1959).

111. F. Kaczmarek and J. Lutomski, Biul. Inst. Przemyslu Zielarski. Poznan., 8, 1 (1962).

112. Z. Szabo, Herba Hung., 2, 99 (1963).

113. E. M. Karácsony, I. Gyenes and Cs. Lörincz, Acta Pharm. Hung., 35, 280 (1965).

114. H. K. Schnoes, A. L. Burlingame and K. Biemann, Tetrahedron Letters, 1962, 993.

115. F. Bartlett, W. I. Taylor and Raymond-Hamet, Compt. Rend., 249, 1259 (1959).

116. M. F. Bartlett, R. Sklar, A. F. Smith and W. I. Taylor, J. Org. Chem., 28, 2197 (1963).

117. W. G. Kump and H. Schmid, Helv. Chim. Acta, 44, 1503 (1961).

118. E. S. Zabolotnaya, A. S. Belikov, S. P. Ivaschenko and M. M. Molodozhnikov, Med. Prom. SSSR, 18 (5) (1964).

119. A. A. Gorman and H. Schmid, Monatsh. Chem., 98, 1554 (1967).

120. R. Kaschnitz and G. Spiteller, Monatsh. Chem., 96, 909 (1965).

121. B. W. Bycroft, D. Schumann, M. B. Patel and H. Schmid, Helv. Chim. Acta, 47, 1147 (1964).

122. K. Biemann, M. Friedmann-Spiteller and G. Spiteller, Tetrahedron Letters, 1961, 485.

123. G. F. Smith and M. A. Wahid, J. Chem. Soc., 1963, 4002.

124. F. Walls, O. Collera and A. Sandoval, Tetrahedron, 2, 173 (1958).

125. H. H. A. Linde, Helv. Chim. Acta, 48, 1822 (1965).

126. P. Relyveld, Pharm. Weekblad., 98, 175 (1963).

127. O. Hesse, Liebigs Ann. Chem., 211, 249 (1882).

128. C. Djerassi, L. N. Antonaccio, H. Budzikiewicz and
     J. M. Wilson, Tetrahedron Letters, 1962, 1001.

129. O. O. Orazi, R. A. Corral, J. S. E. Holker and C.
     Djerassi, J. Org. Chem., 21, 979 (1956).

130. J. Schmutz and H. Lehner, Helv. Chim. Acta, 42, 874
     (1959).

131. L. D. Antonaccio, N. A. Pereira, B. Gilbert, H.
     Vorbrueggen, H. Budzikiewicz, J. M. Wilson, L. J.
     Durham and C. Djerassi, J. Am. Chem. Soc., 84, 2161
     (1962).

132. E. Schlittler and E. Gellért, Helv. Chim. Acta, 34,
     920 (1951).

133. A. Chatterjee, G. R. Ghosal, N. Adityachaudhury and
     S. Ghosal, Chem. Ind. (London), 1961, 1034.

134. A. Walser and C. Djerassi, Helv. Chim. Acta, 48, 391
     (1965).

135. M. P. Cava, S. S. Tjoa, Q. A. Ahmed and A. I. Da-
     Rocha, J. Org. Chem., 33, 1055 (1968).

136. C. Djerassi, H. Budzikiewicz, J. M. Wilson, J. Gosset,
     J. Le Men and M. -M. Janot, Tetrahedron Letters,
     1962, 235.

137. J. Gosset, J. Le Men and M. -M. Janot, Ann. Pharm.
     Franc., 20, 448 (1962).

138. M. Plat, R. Le May, J. Le Men, M. -M. Janot, C.
     Djerassi and H. Budzikiewicz, Bull. Soc. Chim. France
     1965, 2497.

139. M. -M. Janot, J. Le Men and C. Fan, Ann. Pharm.
     Franc., 15, 513 (1957).

140. Sh. Z. Kasymov, P. Kh. Yuldashev and S. Yu. Yunusov, Khim. Prirodn. Soedin. Akad. Nauk Uz. SSR, 1966 (4), 260.

141. P. Kh. Yuldashev and S. Yu. Yunusov, Uzbeksk. Khim. Zh., 1963 (1), 44.

142. S. Yu. Yunusov and P. Kh. Yuldashev, Zh. Obshch. Khim., 27, 2015 (1957).

143. B. Das, K. Biemann, A. Chatterjee, A. B. Ray and P. L. Majumder, Tetrahedron Letters, 1966, 2483.

144. R. M. Bernal, A. Villegas-Castillo and O. P. Espejo, Experientia, 16, 353 (1960).

145. B. P. Korzun, A. F. St. André and P. R. Ulshafer, J. Am. Pharm. Assoc., Sci. Ed., 46, 720 (1957).

146. G. Iacobucci and V. Delofeu, Anales Asoc. Quim. Arg., 46, 143 (1958).

147. W. J. McAleer, R. G. Weston and E. E. Hower, Chem. Ind. (London), 1956, 1387.

148. B. O. G. Schuler and F. L. Warren, J. Chem. Soc., 1956, 215.

149. W. E. Court, W. C. Evans and G. E. Trease, J. Pharm. Pharmacol., 10, 380 (1958).

150. D. A. A. Kidd, Chem. Ind. (London), 1957, 1013.

151. D. A. A. Kidd, J. Chem. Soc., 1958, 2432.

152. D. Banes, A. E. H. Houk and J. Wolff, J. Am. Pharm. Assoc., Sci. Ed., 47, 625 (1958).

153. B. U. Vergara, J. Am. Chem. Soc., 77, 1864 (1955).

154. A. Stoll and A. Hofmann, Soc. Biol. Chemists, India, Silver Jubilee Souvenir, 1955, 248.

155. M. W. Klohs, M. D. Draper, F. Keller and J. Petracek J. Am. Chem. Soc., 76, 7381 (1954).

156. K. Yamaguchi and H. Shoji, Eisei Shikensho Hokoku, <u>76</u>,
      99 (1958).

157. B. O. G. Schüler and F. L. Warren, Chem. Ind.
      (London), <u>1955</u>, 1593.

158. C. K. Atal, J. Am. Pharm. Assoc., Sci. Ed., <u>48</u>, 37
      (1959).

159. A. Chatterjee and S. Talapatra, Naturwissenschaften,
      <u>42</u>, 182 (1955).

160. W. B. Mors, P. Zaltzman, J. J. Beereboom, S. C.
      Pakrashi and C. Djerassi, Chem. Ind. (London), <u>1956</u>,
      173.

161. J. M. Müller, Experientia, <u>13</u>, 479 (1957).

162. G. Dillemann, R. Paris and P. Chaumelle, Ann. Pharm.
      Franc., <u>16</u>, 504 (1958).

163. R. Paris, G. Dilleman and P. Chaumelle, Ann. Pharm.
      Franc., <u>15</u>, 360 (1957).

164. X. Monseur, J. Pharm. Belg., <u>12</u>, 39 (1957).

165. D. S. Rao and S. B. Rao, J. Am. Pharm. Assoc., Sci.
      Ed., <u>44</u>, 253 (1955).

166. A. K. Kiang and A. S. C. Wan, J. Chem. Soc., <u>1960</u>,
      1394.

167. G. Iacobucci and V. Deulofeu, J. Org. Chem., <u>1960</u>,
      1394.

168. S. C. Pakrashi, C. Djerassi, R. Wasicky and N. Neuss,
      J. Am. Chem. Soc., <u>77</u>, 6687 (1955).

169. F. A. Hochstein, J. Am. Chem. Soc., <u>77</u>, 5744 (1955).

170. J. M. Müller, E. Schlittler and H. J. Bein, Experientia,
      <u>8</u>, 338 (1952).

171. H. Kaneko, R. Fujimoto, K. Namba and K. Hayashi,
      Yakugaku Zasshi, <u>82</u>, 1493 (1962).

172. H. Kaneko, R. Fujimoto, K. Namba and K. Hayashi,
      Yakugaku Zasshi, 82, 1489 (1962).

173. H. T. Cardoso and I. A. A. Venancio, Rev. Brasil.
      Biol., 16, 231 (1956).

174. C. Djerassi, J. Fishman, M. Gorman, J. P. Kutney
      and S. C. Pakrashi, J. Am. Chem. Soc., 79, 1217 (1957).

175. C. Djerassi and J. Fishman, Chem. Ind.,(London), 1955,
      627.

176. J. Poisson, A. Le Hir, R. Goutarel and M. -M. Janot,
      Compt. Rend., 238, 1607 (1954).

177. C. Djerassi, M. Gorman, A. L. Nussbaum and J.
      Reynoso, J. Am. Chem. Soc., 76, 4463 (1954).

178. M. Ishidate, M. Okada and K. Saito, Pharm. Bull.
      (Tokyo), 3, 319 (1955).

179. F. A. Hochstein, K. Murai and W. H. Boegemann, J.
      Am. Chem. Soc., 77, 3551 (1955).

180. W. E. Court, F. S. Shakim and A. F. Stewart, Planta
      Med., 15, 282 (1967).

181. W. E. Court, F. S. Shakim and A. F. Stewart, Planta Med.
      15, 173 (1967).

182. W. E. Court, Canadian J. Pharm. Sci., 1, 83 (1966).

183. C. H. Wei, Yao Hseuh Pao, 12, 429 (1965): Chem.
      Abstr., 63, 16779 (1965).

184. E. A. Moreira, Tribuna Farm. (Brazil), 31, 57 (1964).

185. L. Batllori and R. San Martin, Farmacognosia (Madrid),
      24, (1-2), 1 (1964).

186. A. F. St. André, B. Korzun and F. Weinfeldt, J. Org.
      Chem., 21, 480 (1956).

187. J. S. E. Holker, M. Cais, F. A. Hochstein and C.
      Djerassi, J. Org. Chem., 24, 314 (1959).

188. N. K. Basu and B. Sarkar, Nature (London), 181, 552
      (1958).

189. B. N. Nazir and K. I. Handa, J. Sci. Ind. Res. (India),
      18B, 175 (1959).

190. B. C. Bose, R. Vijayvargiya and J. N. Bhatnagar,
      Indian J. Med. Sci., 13, 905 (1959).

191. W. D. Crow and Y. M. Greet, Australian J. Chem., 8,
      461 (1955).

192. G. H. Svoboda, J. Am. Pharm. Assoc., Sci. Ed., 46,
      508 (1957).

193. R. G. Curtis, G. J. Handley and T. C. Somers, Chem.
      Ind. (London), 1955, 1598.

194. T. R. Govindachari, N. Viswanathan, B. R. Pai and T.
      S. Savitri, Tetrahedron, 21, 2951 (1965).

195. B. Das, K. Biemann, A. Chatterjee, A. B. Ray and P.
      L. Majumder, Tetrahedron Letters, 1965, 2239

196. F. A. Doy and B. P. Moore, Australian J. Chem., 15,
      548 (1962).

197. E. G. Pobedimova, Flora SSSR, 18, 646 (1952).

198. L. H. Pammel, "Poisonous Plants of the World", The
      Torch Press, Cedar Rapids, Iowa, 1911.

199. J. M. Watt and M. G. Breyer-Brandwijk, "The Medi-
      cinal and Poisonous Plants of Southern Africa", E. and
      S. Livingstone, Edinburgh, 1932.

200. G. Planchon and E. Collin, "Les Drogues Simples
      d'Origine Vegetale", 1st Ed., F. Savy, Paris, 1875.

201. M. -M. Janot, "Les Alcaloides des Pervenches (Vinca
      minor L. et Vinca major L.)", Hommage au Doyen
      Rene Fabre (Extrait), 1956.

202. A. H. Murphee, Florida Anthropologist, 18, 175 (1965).

203. "The Extra Pharmacopoeia of Martindale", 19th Ed.,
      Vol. I, The Pharmaceutical Press, London, 1928.

204. "The Extra Pharmacopoeia of Martindale", 24th Ed.,
      The Pharmaceutical Press, London, 1958.

205. N. R. Farnsworth, Lloydia, 24, 105 (1961).

206. A. Orechoff, H. Gurewitch and S. Norkina, Arch.
      Pharm. 272, 70 (1934).

207. A. Quevauviller, J. Le Men and M. -M. Janot, Ann.
      Pharm. Franc., 13, 328 (1955).

208. E. H. Lucas, A. Lickfeldt, R. Y. Gottshall and J. C.
      Jennings, Bull. Torrey Botan. Club, 78, 310 (1951).

209. B. J. Abbott, J. L. Hartwell, J. Leiter, R. E. Perdue,
      Jr. and S. A. Schepartz, Cancer Res., 26, 1302 (1966).

210. N. R. Farnsworth, H.H.S. Fong, R. N. Blomster and
      F. J. Draus, J. Pharm. Sci., 51, 217 (1962).

211. H. H. S. Fong, Alkaloid Assay Methods and Hematologi-
      cal Studies of Extracts from Vinca major (L.) var.
      major Pich., M.S. Thesis, University of Pittsburgh,
      1959.

212. H. J. Bein, Pharmacol. Rev., 8, 435 (1956).

213. M. E. Wall, M.M.Krider, C. F. Krewson, C. R. Eddy,
      J. J. Willaman, D. S. Correll and H. S. Gentry, Bull.
      AIC-363, U.S.D.A., Agric. Res. Serv., 1954.

214. E. C. Bate-Smith, Biochem. J., 58, 126 (1954).

215. M. Arnaud, Compt. Rend., 109, 911 (1889).

216. M. Bridel and C. Charaux, Bull. Soc. Chim. Biol., 1,
      40 (1926).

217. Z. Kowalewski and M. Kowalska, Poznan. Tow.
      Przyjaciol Nauk, Wyd. Lek., Pr. Kom. Farm., 5, 55
      (1966).

218. R. K. Ibrahim and G. H. N. Towers, Nature (London), <u>184</u>, 1903 (1959).

219. J. Le Men and H. Pourrat, Ann. Pharm. Franc., <u>11</u>, 449 (1953).

220. J. Trojánek and J. Hodková, Collection Czech. Chem. Commun., <u>27</u>, 2981 (1962).

221. M. -M. Janot and J. Le Men, Compt. Rend., <u>238</u>, 2550 (1954).

222. M. -M. Janot and J. Le Men, Compt. Rend., <u>240</u>, 909 (1955).

223. M. -M. Janot and J. Le Men, Ann. Pharm. Franc., <u>13</u>, 325 (1955).

224. M. -M. Janot and J. Le Men, Compt. Rend., <u>241</u>, 767 (1955).

225. M. -M. Janot, J. Le Men, K. Aghoramurthy and R. Robinson, Experientia, <u>11</u>, 343 (1955).

226. M. -M. Janot, J. Le Men, and Y. Hammouda, Compt. Rend., <u>243</u>, 85 (1956).

227. J. Gosset, J. Le Men, and M. -M. Janot, Bull. Soc. Chim. France, <u>1961</u>, 1033.

228. J. Gosset-Garnier, J. Le Men and M. -M. Janot, Bull. Soc. Chim. France, <u>1965</u>, 676.

229. N. Abdurakhimova, P. Kh. Yuldashev and S. Yu. Yunusov, Dokl. Akad. Nauk SSSR, <u>21</u> (4), 33 (1964).

230. N. Abdurakhimova, P. Kh. Yuldashev and S. Yu. Yunusov, Khim. Prirodn. Soedin. Akad. Nauk Uz. SSR, <u>1965</u> (3), 224.

231. I. Ognyanov, B. Pyuskyulev, I. Kompis, T. Sticzay, G. Spiteller, M. Shamma and R. J. Shine, Tetrahedron, <u>24</u>, 4641 (1968).

232.   P. Potier, R. Beugelmans, J. Le Men and M. -M.
       Janot, Ann. Pharm. Franc., 23, 61 (1965).

233.   M. Plat, R. Lemay, J. Le Men, M. -M. Janot, C.
       Djerassi and H. Budzikiewicz, Bull. Soc. Chim. France
       1965, 2497.

234.   J. L. Kaul and J. Trojánek, Lloydia, 29, 26 (1966).

235.   J. L. Kaul and J. Trojánek, Sci. Pharm., 1, 325 (1966).

236.   P. Kh. Yuldashev, J. L. Kaul, Z. Kablicova, J.
       Trojánek and S. Yu. Yunusov, Khim. Prirodn. Soedin.
       Akad. Nauk Uz. SSR, 1966, 19?.

237.   B. Borkowski, E. Batkiewicz and K. Drost, Disserta-
       tiones Pharm., 16, 171 (1964).

238.   M. B. Patel, J. Poisson, J. L. Pouset and J. M.
       Rowson, J. Pharm. Pharmacol. 17, 323 (1965).

239.   S. Goodwin and E. C. Horning, Chem. Ind. (London),
       1956, 846.

240.   Sh. Z. Kasymov, P. Kh. Yuldashev and S. Yu. Yunusov,
       Dokl. Akad. Nauk SSSR, 162, 102 (1965).

241.   T. A. Henry, J. Chem. Soc., 1932, 2759.

242.   R. Paris and H. Moyse-Mignon, Compt. Rend., 236,
       1993 (1953).

243.   M. -M. Janot and J. Le Men, Compt. Rend., 243, 1789
       (1956).

244.   I. Ognyanov, P. Dalev, H. Dutschevska and N. Mollov,
       Compt. Rend. Acad. Bulgare Sci., 17, 153 (1964).

245.   S. Yu. Yunusov and P. Kh. Yuldashev, Dokl. Akad.
       Nauk Uz. SSR, 1956 (9), 23.

246.   J. Keck, Naturwissenschaften, 42, 391 (1955).

247.   D. K. Atal, J. Am. Pharm. Assoc., Sci. Ed., 48, 37
       (1959).

248. S. Bhattacharji, M. M. Dhar and M. L. Dhar, J. Sci.
      Ind. Res. (India), 21B, 454 (1962).

249. R. Pernet, J. Philippe and G. Combes, Ann. Pharm.
      Franc., 20, 527 (1962).

250. R. Salkin, N. Hosansku and R. Jaret, J. Am. Pharm.
      Assoc., Sci. Ed., 50, 1038 (1961).

251. E. Haack, A. Popelack, H. Spingler and F. Kaiser,
      Naturwissenschaften, 41, 214 (1954).

252. E. Schlittler, H. Sauer and J. M. Müller, Experientia,
      10, 133 (1954).

253. B. Gilbert, A. P. Duarte, Y. Nakagawa, J. A. Joule, S. E.
      Flores, J. A. Brissolesse, J. Campello, E. P. Carrazzoni,
      R. J. Owellen, E. C. Blossey, K. S. Brown, Jr. and C.
      Djerassi, Tetrahedron, 21, 1141 (1965).

254. B. Gilbert, J. A. Brissolese, N. Finch, W. I. Taylor,
      H. Budzikiewicz, J. M. Wilson, and C. Djerassi, J. Am.
      Chem. Soc., 85, 1523 (1963).

255. R. R. Arndt, S. H. Brown, N. C. Ling, P. Roller, C.
      Djerassi, J. M. Ferreira, F., B. Gilbert, E. C. Miranda,
      S. E. Flores, A. P. Duarte and E. P. Carrazzoni,
      Phytochemistry 6, 1653 (1967).

256. M. B. Patel, J. Poisson, J. L. Pousset, and J. M.
      Rowson, J. Pharm. Pharmacol., 16, 163T (1964).

257. A. G. Kurmukov, Farmakol. i Toksikol., 31, 47 (1968).

258. J. D. Achelis and G. Kroneberg, Naturwissenschaften,
      40, 625 (1953).

259. G. Kroneberg and J. D. Achelis, Arzneimittel-Forsch.,
      4, 270 (1954).

260. J. Gosset, J. Le Men and M. -M. Janot, Ann. Pharm.
      Franc., 20, 448 (1962).

261. M. Falco, J. Gosset-Garnier, E. Fellion and J. Le Men,
      Ann. Pharm. Franc., 22, 455 (1964).

262. M. -M. Janot, J. Le Men, J. Gosset and J. Levy, Bull.
      Soc. Chim. France, 1962, 1079.

263. M. Ohashi, H. Budzikiewicz, J. M. Wilson, C. Djerassi,
      J. Levy, J. Gosset, J. Le Men and M. -M. Janot, Tetra-
      hedron, 19, 2241 (1963).

264. S. Silvers and A. Tulinsky, Tetrahedron Letters, 1962, 339.

265. C. Djerassi, H. Budzikiewicz, J. M. Wilson, J. Gosset,
      J. Le Men and M. -M. Janot, Tetrahedron Letters, 1962,
      235.

266. B. C. Das, J. Garnier-Gosset, J. Le Men and M. -M.
      Janot, Bull. Soc. Chim. France, 1965, 1903.

267. S. Bose, S. K. Talapatra and A. Chatterjee, J. Indian
      Chem. Soc., 33, 379 (1956).

268. M. Ishidate, M. Okada and K. Saito, Pharm. Bull.
      (Tokyo), 3, 319 (1955).

269. K. Bodendorf and H. Eder, Naturwissenschaften, 40,
      342 (1953).

270. A. Stoll, and A. Hofmann, Helv. Chim. Acta, 36, 1143
      (1953).

271. J. Poisson and R. Goutarel, Bull. Soc. Chim. France,
      1956, 1703.

272. V. M. Malikov, P. Kh. Yuldashev and S. Yu. Yunusov,
      Khim. Prirodn. Soedin. Akad. Nauk Uz. SSR, 1966,
      (5), 338.

273. S. Markey, K. Biemann and B. Witkop, Tetrahedron
      Letters, 1967, 157.

274. H. Rapoport and R. E. Moore, J. Org. Chem., 27,
      2981 (1962).

275. H. R. Arthur, S. R. Johns, J. R. Lamberton and S. N.
      Loo, Australian J. Chem., 21, 1399 (1968).

276. S. Ya. Zolotnitskaya, I. S. Melkumyan and V. E. Vaskanyan, Izv. Akad. Nauk Arm. SSR., Biol. Nauki, 15, 33 (1962).

277. K. S. Roussinov, D. Zhelyaskov and V. Georgiev, Arch. Ital. Sci. Farmacol., 11, 83 (1961).

278. K. S. Roussinov, D. Zhelyaskov and V. Georgiev, Izv. Inst. po Fiziol. Bulgar. Akad. Nauk, 5, 271 (1962).

279. K. S. Roussinov, D. Zhelyaskov and V. Georgiev, Compt. Rend. Acad. Bulgare Sci., 15, 329 (1962).

280. P. Panov, I. Ognyanov, N. Mollov, K. S. Roussinov, V. Georgiev and D. Zhelyaskov, Compt. Rend. Acad. Bulgare Sci., 13, 39 (1961).

281. N. R. Farnsworth, R. N. Blomster, and J. P. Buckley, J. Pharm. Sci., 56, 23 (1967).

282. D. A. Bocharova, Aptechn. Delo, 8, 23 (1959).

283. I. Ognyanov and B. Pyuskyulev, Chem. Ber., 99, 1008 (1966).

284. I. Ognyanov, B. Pyuskyulev and G. Spiteller, Monatsh. Chem., 97, 857 (1966).

285. N. Mollov, J. Mokrý, I. Ognyanov and P. Dalev, Compt. Rend. Acad. Bulgare Sci., 14, 43 (1961).

286. I. Ognyanov, B. Pyuskyulev, M. Shamma, J. A. Weiss and R. J. Shine, Chem. Commun., 1967, 579.

287. I. Ognyanov, Chem. Ber., 99, 2052 (1966).

288. B. Pyuskyulev, I. Kompiš, I. Ognyanov and G. Spiteller, Collection Czech. Chem. Commun., 32, 1289 (1967).

289. B. Pyuskyulev, I. Ognyanov and P. Panov, Tetrahedron Letters, 1967, 4559.

290. I. Ognyanov, B. Pyuskyulev, B. Bozjanov and M. Hesse, Helv. Chim. Acta, 50, 754 (1967).

291. E. S. Zabolotnaya and E. V. Bukreeva, Zh. Obshch. Khim., 33, 3780 (1963).

292. B. K. Moza and J. Trojánek, Chem. Ind. (London), 1962, 1425.

293. B. Moza, J. Trojánek, A. K. Bose, K. G. Das and P. Funke, Lloydia, 27, 416 (1964).

294. E. M. Maloney, N. R. Farnsworth, R. N. Blomster, D. J. Abraham and A. G. Sharkey, Jr., J. Pharm. Sci., 54, 1166 (1965).

295. M. Tin-Wa, H. H. S. Fong, R. N. Blomster and N. R. Farnsworth, J. Pharm. Sci. In press.

296. H. Tomczyk, Dissertationes Pharm. Pharmacol., 20, 63 (1968).

297. B. Das, E. Fellion and M. Plat, Compt. Rend., 264, 1765 (1967).

298. M. -M. Janot, H. Pourrat and J. Le Men, Bull. Soc. Chim. France, 1954, 707.

299. O. Collera, F. Walls, A. Sandoval, F. Garciá, J. Herrán and M. C. Perezamador, Bol. Inst. Quim. Nal. Auton. Mex., 14, 3 (1962).

300. Kh. N. Aripov, T. T. Shakirov, and P. Kh. Yuldashev, Khim. Prirodn. Soedin. Akad. Nauk Uz. SSR, 1966 (4), 293.

301. D. Stauffacher, Helv. Chim. Acta, 44, 2006 (1961).

302. M. B. Sultanov and T. Saidkasymov, Dokl. Akad. Nauk Uz. SSR, 1965 (6), 41.

303. M. B. Sultanov and T. Saidkasymov, Farmakol. Alkaloidov, Akad. Nauk Uz. SSR, Inst. Khim. Rast. Veshchestv, 1965 (2), 104.

304. T. Saidkasymov and M. B. Sultanov, Vopr. Ispol'z. Mineral'm i Rast.' Syr'ya Srednei Azii, Akad. Nauk Uz. SSR, Otd. Geol. -Khim. Nauk, 1961 (13) 176.

305. T. Saidkasymov, Vopr. Biol. i Kraevoi Med., Akad.
     Nauk Uz. SSR, Otd. Biol. Nauk, 1960, 242.

306. M.B. Sultanov and E. B. Bailiekov, Ref. Zh. Otd.
     Vypusk. Farmakol. i Toksikol., 54, 121 (1962).

307. M. B. Sultanov and T. A. Egorova, Dokl. Akad. Nauk
     Uz. SSR, 1961 (10) 28.

308. M. B. Sultanov, Isv. Akad. Nauk Uz. SSR, Ser. Med.,
     1959 (3), 38.

309. A. A. Vakhabov and M. B. Sultanov, Akad. Nauk Uz.
     SSR, Khim.-Tekh. i Biol. Otd., 1966, 30.

310. A. G. Kurmukov and M. B. Sultanov, Uzbeksk. Biol.
     Zh., 11 (1), 32 (1967).

311. A. G. Kurmukov, Farmakol. i Toksikol., 30 (3), 286
     (1967).

312. A. A. Vakhabov and M. B. Sultanov, Farmakol.
     Alkaloidov, Akad. Nauk Uz. SSR, Inst. Khim. Rast.'
     Veshchestv, 1965 (2), 183.

313. M. B. Sultanov and A. G. Kurmukov, Farmakol.
     Alkaloidov, Akad. Nauk Uz. SSR, Inst. Khim. Rast.'
     Veshchestv, 1965 (2), 128.

314. A. G. Kurmukov, L. N. Goldberg, M. B. Sultanov, L.
     A. Vysotskaya and A. A. Kulikov, Ref. Zh. Otd. Vypusk.
     Farmakol. i Khimioter. Sredstva Toksikol., 1967,
     3/54/490.

315. A. G. Kurmukov, Farmakol. i Toksikol., 31, 47 (   ).

316. A. G. Kurmukov, Med. Zh. Uzbeksk., 1967 (6), 49.

317. A. G. Kurmukov and M. B. Sultanov, Akad. Nauk Uz.
     SSR, Khim.-Tekh. i Biol. Otd., 1966, 26.

318. A. A. Vakhabov and M. B. Sultanov, Akad. Nauk. Uz.
     SSR, Khim.-Tekh. i Biol. Otd., 1966, 30.

319. A. A. Abidov, S. M. Muchamedov and A. A. Abdychaku-
     mov, Med. Zh. Uzbekistana, 1963, 45.

320. D. A. Bocharova, Rast. Resur., 2 (4), 513 (1966).

321. N. I. Koretskaya and L. M. Utkin, Zh. Obshch. Khim.,
     33, 2065 (1963).

322. V. M. Malikov and S. Yu. Yunusov, Khim. Prirodn.
     Soedin. Akad. Nauk Uz. SSR, 1967 (2), 142.

323. S. Yu. Yunusov and P. Kh. Yuldashev, Zh. Obshch.
     Khim., 27, 2072 (1957).

324. P. Kh. Yuldashev and S. Yu. Yunusov, Dokl. Akad.
     Nauk Uz. SSR, 1960, (3), 28.

325. N. Abdurakhimova, P. Kh. Yuldashev and S. Yu.
     Yunusov, Dokl. Akad. Nauk Uz. SSR, 1964, (2), 29.

326. Sh. Z. Kasymov, P. Kh. Yuldashev and S. Yu. Yunusov,
     Dokl. Akad. Nauk SSSR, 163, 1400 (1965).

327. V. M. Malikov, P. Kh. Yuldashev and S. Yu. Yunusov,
     Khim. Prirodn. Soedin. Akad. Nauk Uz. SSR, 1966 (5),
     338.

328. M. A. Kuchenkova, P. Kh. Yuldashev and S. Yu.
     Yunusov, Dokl. Akad. Nauk Uz. SSR, 1964, (11), 42.

329. M. A. Kuchenkova, P. Kh. Yuldashev and S. Yu.
     Yunusov, Izv. Akad. Nauk SSSR, Ser. Khim., 1965 (12),
     2152.

330. P. Kh. Yuldashev, Kh. Ybaev, M. A. Kuchenkova and S.
     Yu. Yunusov, Khim. Prirodn. Soedin. Akad. Nauk Uz.
     SSR, 1965 (1), 34.

331. N. Abdurakhimova, P. Kh. Yuldashev and S. Yu.
     Yunusov, Dokl. Akad. Nauk SSSR, 173, 87 (1967).

332. N. Abdurakhimova, P. Kh. Yuldashev and S. Yu.
     Yunusov, Khim. Prirodn. Soedin. Akad. Nauk Uz. SSR,
     1967, (5), 310.

333. Kh. Y. Ybaev, P. Kh. Yuldashev and S. Yu. Yunusov,
     Dokl. Akad. Nauk Uz. SSR, 1964, (10), 34.

334. Kh. Y. Ybaev, P. Kh. Yuldashev and S. Yu. Yunusov,
     Izv. Akad. Nauk SSSR, Ser. Khim., 1965 (11), 1992.

335. D. A. Pachymov, V. N. Malikov and S. Yu. Yunusov,
     Khim. Prirodn. Soedin. Akad. Nauk Uz. SSR, 1967 (5),
     354.

336. M. A. Kuchenkova, P. Kh. Yuldashev and S. Yu.
     Yunusov, Khim. Prirodn. Soedin. Akad. Nauk Uz. SSR,
     1967 (1), 65.

337. P. Kh. Yuldashev, V. M. Malikov and S. Yu. Yunusov,
     Dokl. Akad. Nauk Uz. SSR, 1960 (1), 25.

338. V. M. Malikov, P. Kh. Yuldashev and S. Yu. Yunusov,
     Dokl. Akad. Nauk Uz. SSR, 1963 (4), 21.

339. P. Kh. Yuldashev and S. Yu. Yunusov, Dokl. Akad.
     Nauk SSSR, 154, 1412 (1964).

340. P. Kh. Yuldashev and S. Yu. Yunusov, Khim. Prirodn.
     Soedin. Akad. Nauk Uz. SSR, 1965 (2), 110.

341. P. Kh. Yuldashev and S. Yu. Yunusov, Dokl. Akad. Nauk
     SSR, 163, 123 (1965).

342. Sh. Z. Kasymov, H. N. Aripov, T. T. Shakirov and
     S. Yu. Yunusov, Khim. Prirodn. Soedin. Akad. Nauk
     Uz. SSR, 1967 (5), 352.

343. T. R. Govindachari, N. Viswanathan, B. R. Pai and
     T. S. Savitri, Tetrahedron, 21, 2951 (1965).

344. W. D. Crow and M. Michael, Australian J. Chem., 8,
     129 (1955).

345. C. Kump and H. Schmid, Helv. Chim. Acta, 45, 1090
     (1962).

346. W. G. Kump, M. B. Patel, J. W. Rowson and H.
     Schmid, Helv. Chim. Acta, 47, 1497 (1964).

347. A. K. Kiang, H. Lee, J. Goh and A. S. C. Wan,
      Lloydia, 27, 220 (1964).

348. N. Abdurakhimova, Sh. Z. Kasymov and S. Yu. Yunusov,
      Khim. Prirodn. Soedin. Akad. Nauk Uz. SSR, 1968 (2),
      135.

349. H. Szostak and Z. Kowalewski, Z. Herba Pol. 15 (1), 66
      (1969).

350. W. Döpke and H. Meisel, Tetrahedron Letters, 1969 (21),
      1701.

351. W. Döpke and H. Meisel, Pharmazie, 23, 521 (1968).

352. W. Döpke, H. Meisel and G. Spiteller, Tetrahedron Let-
      ters, 1968 (58), 6065.

353. W. Döpke and H. Meisel, Tetrahedron Letters 1970 (10),
      749.

354. N. A. Babaev, Azerb. Med. Zh., 1968 (8), 32.

355. A. M. Aliev and N. A. Babaev, Farmatsiya (Moscow), 18
      (5), 28 (1969).

356. A. M. Aliev and N. A. Babaev, Farmatsiya, 17 (4), 23
      (1968).

357. E. S. Zabolotnaya, E. V. Bukreeva and T. V. Lazur'evskii,
      Khim. -Farm. Zh. 3(1), 32 (1969).

358. N. Abdurakhimova, Sh. Z. Kasymov and S. Yu. Yunusov,
      Khim. Prirodn. Soedin. 1968 (2), 135.

359. D. A. Rakhimov, V. M. Malikov and S. Yu. Yunusov, Khim.
      Prir. Soedin. 5 (4), 332 (1969).

360. D. A. Rakhimov, V. M. Malikov and S. Yu. Yunusov, Khim.
      Prir. Soedin. 5 (5), 461 (1969).

361. V. M. Malikov and S. Yu. Yunusov, Khim. Prir. Soedin.
      5 (1), 65 (1969).

362. D. A. Rakhimov, B. M. Malikov and S. Yu. Yunusov, Khim.
      Prir. Soed. 1969 (4), 330.

363. D. A. Rakhimov, B. M. Malikov and S. Yu. Yunusov, Khim.
      Prir. Soed. 1969 (4), 336.

CHART 1

Eburnamine-type alkaloids isolated from Vinca species.

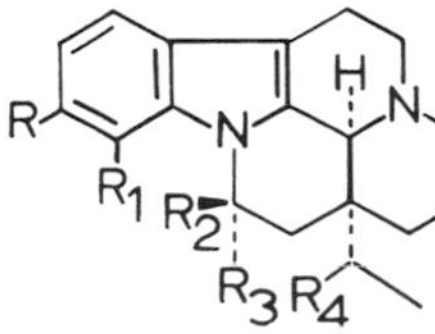

Number	R	R_1	R_2	R_3	R_4	Alkaloid
I	H	H	OH	COOCH$_3$	H$_2$	Vincamine
II	H	H	COOCH$_3$	OH	H$_2$	14-Epivincamine
III	CH$_3$O	H	OH	COOCH$_3$	H$_2$	Vincine
IV	H	H	OH	COOCH$_3$	OH	20-Hydroxyvinca-mine
V	H	H	OH	COOCH$_3$	O=	Vincaminine (Vincareine)
VI	CH$_3$O	H	OH	COOCH$_3$	O=	Vincinine
VII$_{dl}$	H	H		O=		H$_2$	(±)-Eburnamonine (Vincanorine)
VII$_l$	H	H		O=		H$_2$	(-)-Eburnamonine
VIII	CH$_3$O	H		O=		H$_2$	11-Methoxyeburn-amonine
IX	CH$_3$O	CH$_3$O		O=		H$_2$	11, 12-Dimethoxy-eburnamonine

CHART 2

Eburnamine-type alkaloids isolated from <u>Vinca</u> species.

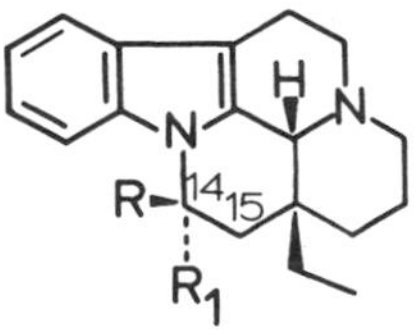

Number	R	R_1	Other	Alkaloid
X	H	OH	-	(-)-Eburnamine
XI	OH	H	-	(+)-Isoeburnamine
XII	-	-	$\Delta^{14,15}$	Eburnamenine

CHART 3

Quebrachamine-type alkaloids isolated from <u>Vinca</u> species.

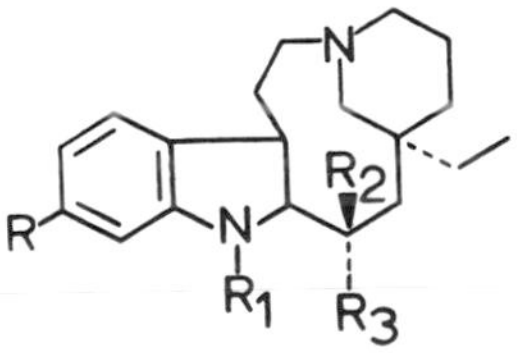

Number	R	R_1	R_2	R_3	Alkaloid
XIII	H	CH_3	$COOCH_3$	H	Vincaminorine
XIV	H	CH_3	H	$COOCH_3$	Vincaminoreine
XV	CH_3O	CH_3	$COOCH_3$	H	Vincaminoridine
XVI	H	H	H	$COOCH_3$	Vincadine
XVII	H	CH_3	H	H	(±)-Ind-N-methyl-quebrachamine
XVIII	H	H	H	H	(+)-Quebrachamine

CHART 4

Aspidosperma-type alkaloids isolated from <u>Vinca</u> species.

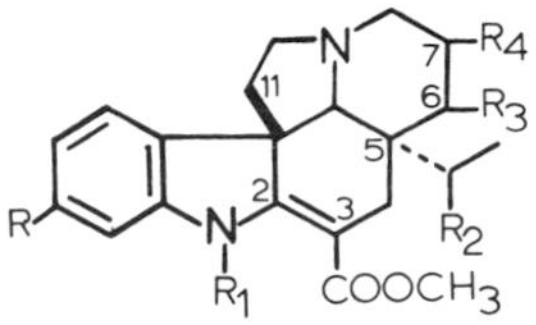

Number	R	R_1	R_2	R_3	R_4	Other	Alkaloid
XIX_{dl}	H	H	H_2	H_2	H_2	-	(±)-Vincadifformine
XIX_1	H	H	H_2	H_2	H_2	-	(-)-Vincadifformine
XX	H	CH_3	H_2	H_2	H_2	-	(±)-Minovine
XXI	H	H	O=	H_2	H_2	-	Minovincine (Minoricine)
XXII	CH_3O	H	O=	H_2	H_2	-	16-Methoxyminovincine (Minoriceine)
XXIII	H	H	OH	H_2	H_2	-	Minovincinine
XXIV	CH_3O	H	H_2		-O-		-	Lochnerinine
XXV	H	H	H_2	H	H	$\Delta^{6,7}$	(-)-Tabersonine
XXVI	CH_3O	H	H_2	H	H	$\Delta^{6,7}$	16-Methoxytabersonine
XXVII	H	H	H_2	H_2	H_2	-	Ervamine[a]
XXVIII	H	H	-	H_2	H_2	2,3-dihydro Ψ-Kopsinine +5-desethyl +5,11-CH(CH$_3$)- +...COOCH$_3$	

[a] Stereochemistry at C_5 is not clear.

CHART 5

Aspidosperma-type alkaloids isolated from <u>Vinca</u> species.

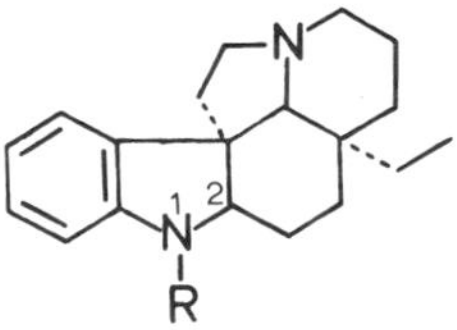

Number	R	Other	Alkaloid
XXIX	CH_3	-	(+)-N-Methylaspidospermi-dine
XXX	-	$\Delta^{1,2}$	(+)-1,2-Dehydroaspido-spermidine

CHART 6

Aspidosperma-type alkaloids isolated from <u>Vinca</u> species.

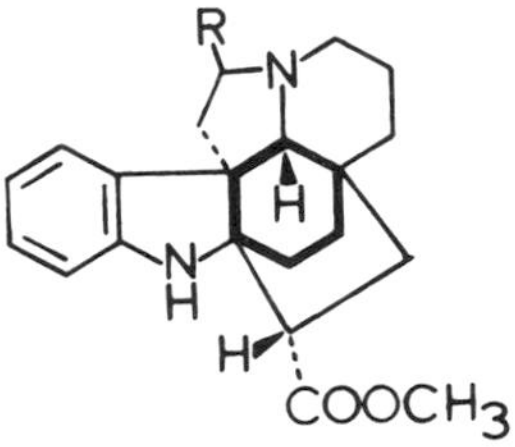

Number	Rxm	Alkaloid
XXXI	H_2	Kopsinine (Erectine)
XXXII	O=	Kopsinilam

CHART 7

Aspidospermatine-type alkaloids isolated from Vinca species.

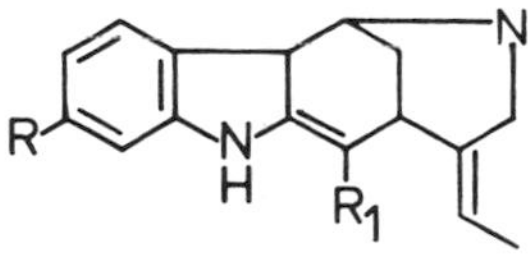

Number	R	R$_1$	Alkaloid
XXXIII	H	CHO	Norfluorocurarine (Vincanine)
XXXIV	OH	COOCH$_3$	Vinervine
XXXV	CH$_3$O	COOCH$_3$	Vinervinine
XXXVI	OH	CHO	Vincanidine

CHART 8

Ajmaline-type alkaloids isolated from Vinca species.

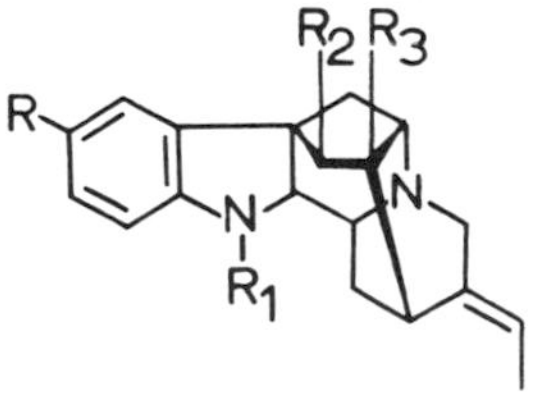

Number	R	R$_1$	R$_2$	R$_3$	Alkaloid
XXXVII	H	CH$_3$	OH	COOCH$_3$	Vincamajine
XXXVIII	H	CH$_3$	CH$_3$OCO	COOCH$_3$	Vincamedine
XXXIX	CH$_3$O	CH$_3$	OH	H	Vincamajoreine
XL	CH$_3$O	CH$_3$	CH$_3$OCO	H	Majoridine
XLI	H	H	OH	COOCH$_3$	Vincarine

CHART 9

Sarpagine-type alkaloids isolated from <u>Vinca</u> species.

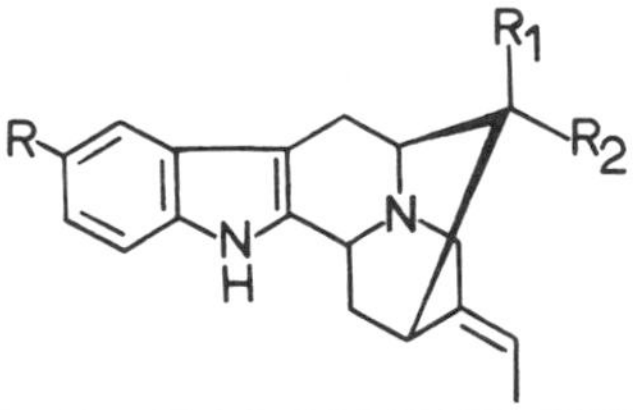

Number	R	R_1	R_2	Alkaloid
XLII	CH_3O	H	CHO	10-Methoxyvellosimine
XLIII	H	H	CHO	Vellosimine
XLIV	OH	H	CH_2OH	Sarpagine
XLV	H	$COOCH_3$	CH_2OH	Akuammidine (Ervamidine)
XLVI	H	H	CH_2OH	Normacusine B (Tombozine)

CHART 10

Sarpagine related-type alkaloids isolated from <u>Vinca</u> species.

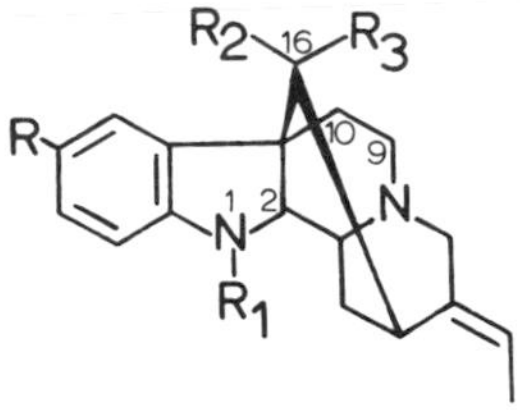

Number	R	R_1	R_2	R_3	Other	Alkaloids
XLVII	H	-	H	$COOCH_3$	$\Delta^{1,2}$	Vincamidine (Strictamine)
XLVIII	H	CH_3	H	$COOCH_3$	2,9(..0..)	Ervincine
XLIX	H	H	CH_3OOC	CH_2OH	-	Vincaridine
L	OH	CH_3	2,16 $(O-CH_2-)$	$COOCH_3$	-	Akuammine (Vincamajoridine)

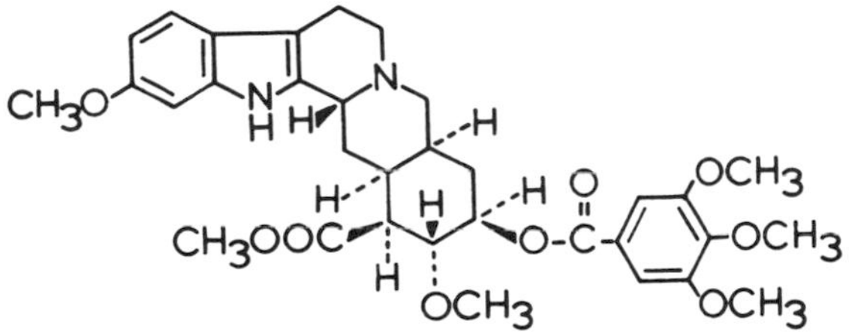

Reserpine LI

CHART 11

Yohimbine-type alkaloids isolated from Vinca species.

CHART 12

Oxindole-type alkaloids isolated from Vinca species.

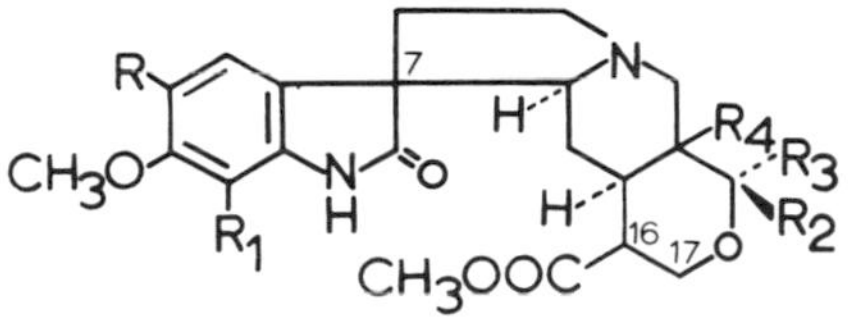

Number	R	R_1	R_2	R_3	R_4	Other	Alkaloid
LII	H	CH_3O	H	CH_3	..H	$\Delta^{16,17}$	Majdine
LIII	CH_3O	H	H	CH_3	..H	$\Delta^{16,17}$	Carapanaubine (Vinine)
LIV	H	CH_3O	H	CH_3	..H	$\Delta^{16,17}$	Isomajdine[a]
LV	CH_3O	H	H	CH_3	►H	-	Herbaline
LVI	H	H	~H	~CH_3	~H	$\Delta^{16,17}$	Vinerine
LVII	H	H	~H	~CH_3	~H	$\Delta^{16,17}$	Vineridine[b]
LVIII	H	H	H	CH_3	..H	$\Delta^{16,17}$	Ericinine[c]

[a]Epimeric with LII at C_7; [b]Epimeric with LVI at C_7;
[c]Stereochemistry at C_3 and C_7 not established.

CHART 13

Yohimbine ring E heterocycle-type alkaloids isolated from

Vinca species.

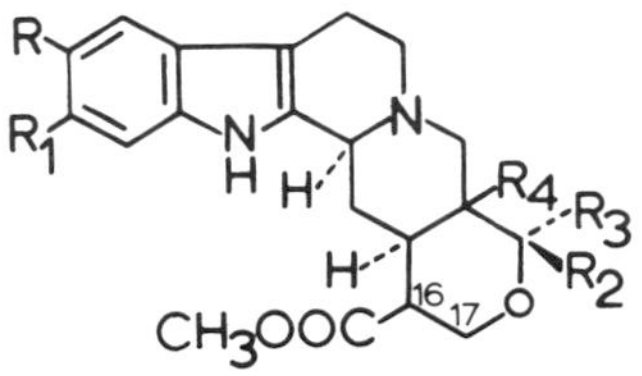

Number	R	R_1	R_2	R_3	R_4	Other	Alkaloid
LIX	H	CH_3O	H	CH_3	..H	$\Delta^{16,\ 17}$	Reserpinine
LX	H	CH_3O	H	CH_3	>H	-	Herbaine
LXI	CH_3O	CH_3O	H	CH_3	>H	-	Herbaceine
LXII	H	H	CH_3	H	..H	-	Ervine

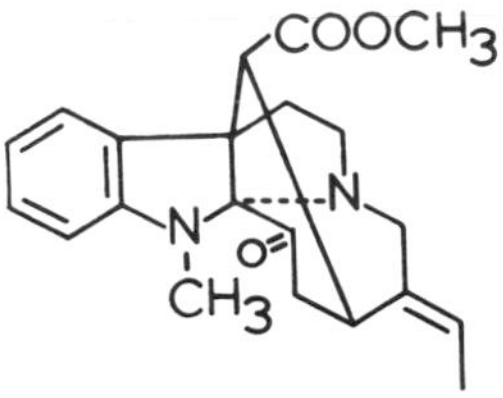

Vincoridine LXIII

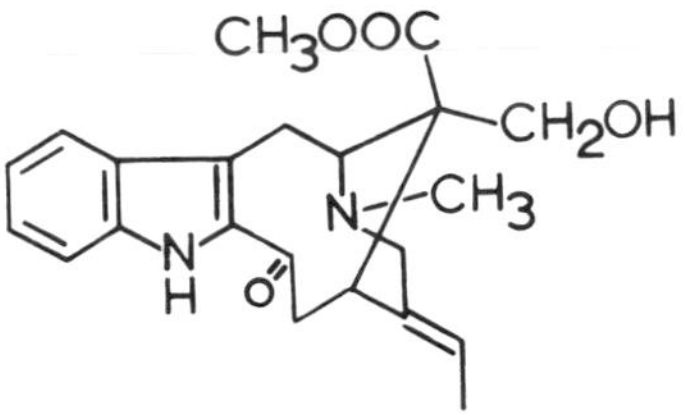

Vincadiffine LXIV

CHART 14

2-Acylindole alkaloids isolated from Vinca species.

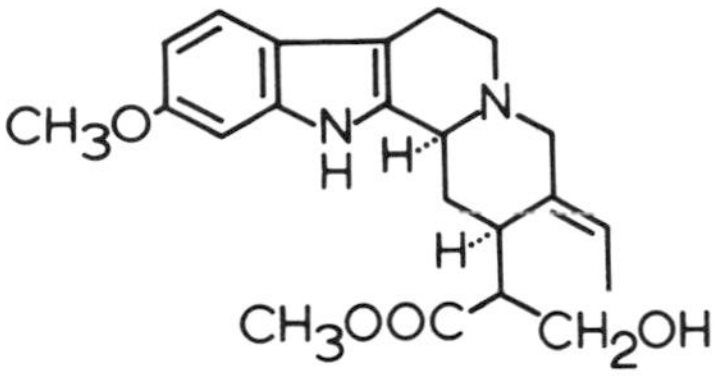

Hervine LXV

CHART 15

Yohimbine ring E seco-type alkaloid isolated from Vinca species.

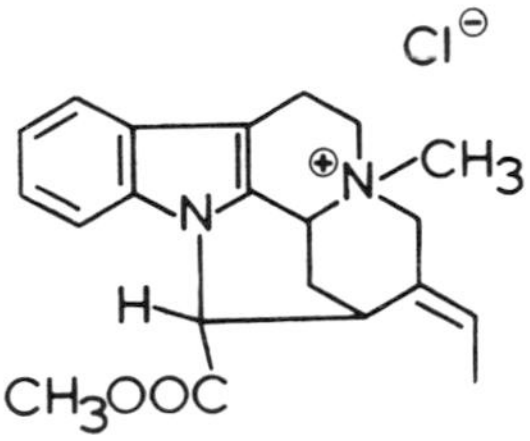

Pleiocarpamine Chloride   LXVI

CHART 16

CHAPTER 3

THE CHEMISTRY OF THE VINCA ALKALOIDS

William I. Taylor

International Flavors & Fragrance
(IFF-R&D)
Union Beach, New Jersey

I. INTRODUCTION

In the preceding chapters botanical, historical and phytochemical aspects of the genus Vinca have been discussed. In
this chapter, the chemistry and the structural elucidation of
the Vinca alkaloids as well as their relationship to other
complex indole alkaloids will be discussed.

The indole alkaloids can be divided into three major groups
(Chart 1), Type I, Type II and Type III bases, which differ in
the way three different arrangements of a 10C fragment have
been fused to a trypamine residue. No bases derived from
Type II precursor have been isolated so far from Vinca species.
Parenthetically, this is one of the outstanding differences
between the alkaloids of Vinca and Catharanthus plants. The
alkaloids so far isolated from V. minor are quite different

from the other well-investigated Vinca species. In fact, with

the exception of strictamine, vincoridine and possibly reser-

pine, no other Type I bases have been found in V. minor. The

remaining alkaloids are Type III bases related to vincamine,

quebrachamine and aspidospermine with the added distinction

for complex natural products that some are found in the race-

mic form. The alkaloids of V. major, V. difformis and V. pub-

escens are largely Type I bases belonging to classes such as

the ajmaline-sarpagine, the yohimbinoid ring E oxygen hetero-

cycles and the derived oxindoles. The alkaloids of V. erecta

and V. herbacea contain yet another variant of the Type I sys-

tem in having akuammicine-like bases along with oxindoles. V.

erecta contains, in addition some kopsine derivatives which

seem to be unusual from this source.

With regard to the alkaloid isolations, it is seldom that all

the bases described from a given Vinca species have been ob-

tained by the same investigators, from the same collection at

the same time. Reserpine, for example, has been reported

from V. minor in Russia, but this compound has not yet been

found by other chemists working in Europe, or North America.

Also, one cannot help wondering whether the plant collections

for V. erecta were botanically homogeneous. Exact answers

to these questions cannot be given until a complete examination of the alkaloids of an individual plant is carried out.

For the most part, the determination of the structures of

the Vinca alkaloids has been greatly facilitated by the wealth

of information, both chemical and physical, available concern-

ing indole alkaloids from other apocynaceous plants. In a num-

ber of cases, a simple transformation was sufficient to solve

the structural problem; in other cases, the molecular archi-

tecture could be deduced from an analysis of the information

provided by various spectrometric determinations. The single

most useful machine has been the mass spectrometer. The

present-day natural product chemist is solving structures with

minute amounts of material. He knows the structure but can

only guess at the chemistry. In the past much new chemistry

was created during efforts to degrade a natural product. To-

day much new chemistry is being discovered by the groups try-

ing to synthesize or biosynthesize these complex heterocycles.

Despite the wealth of practical and theoretical chemistry

brought to bear on these problems, much remains to be

learned.

Table 1 lists the important Vinca alkaloids and their

synonyms. We shall use the name which is deemed to be most

useful for our purposes rather than the one which may have

TABLE 1

Some Vinca Alkaloids and Synonyms

Alkaloids	Synonym
10-Hydroxy-Akuammicine	Vinervine
Akuammidine	Ervamidine
Akuammine	Vincamajoridine
1, 2-Dehydro-Aspidospermidine	Alkaloid 280A
Carapanaubine	Vinine, majoroxine
Eburnamine	Pleiocarpinidine
Herbaceine	Vincaherbinine
Herbaine	Vincaherbine (?)
Hervine	11-Methoxyisositsirikine
Kopsinine	Erectine
Lochnerinine	11-Methoxylochnericine
Norfluorocurarine	Vincanine
11-Hydroxy-Norfluorocurarine	Vincanidine
Majdine	Majoroxine, Alkaloid A5
Iso-Majdine	Alkaloid A4
Majoridine	17-O-Acetyl-10-Methoxytetra-phyllicine
Minoriceine	(-)-11-Methoxy-19-Oxovinca-difformine

Minovincine	Minoricine, base VM-15 (-)-19-Oxovincadifformine
Minovincinine	(-)-19-Hydroxyvincadifformine
Minovine	(±)-1-Methylvincadifformine
Reserpinine	Pubescine
Strictamine	Vincamidine
Tetraphyllicine	"Unnamed base"
Vellosimine	Alkaloid Y
Vincadine	16-Methoxycarbonyl- quebrachamine
Vincamedine	O-Acetylvincamajine
Vincamine	Minorine
16-epi-Vincamine	Isovincamine
11-Methoxy-Vincamine	Vincine
19-Oxo-11-Methoxy-Vincamine	Vincinine
19-Oxo-Vincamine	Vincaminine, Vincareine
Vincaminoreine	1-Methyl-16-Methoxycarbonyl- quebrachamine
Vincaminorine	1-Methyl-16-epi-Methoxy- carbonyl-quebrachamine
Vincarine	17(?)-epivincamajine

priority. For a complete listing of all alkaloids and

structures, see Chapter 2, Charts 1-16.

The numbering system used throughout this chapter

assigns to each atom, see Scheme 1, the same numbers as their

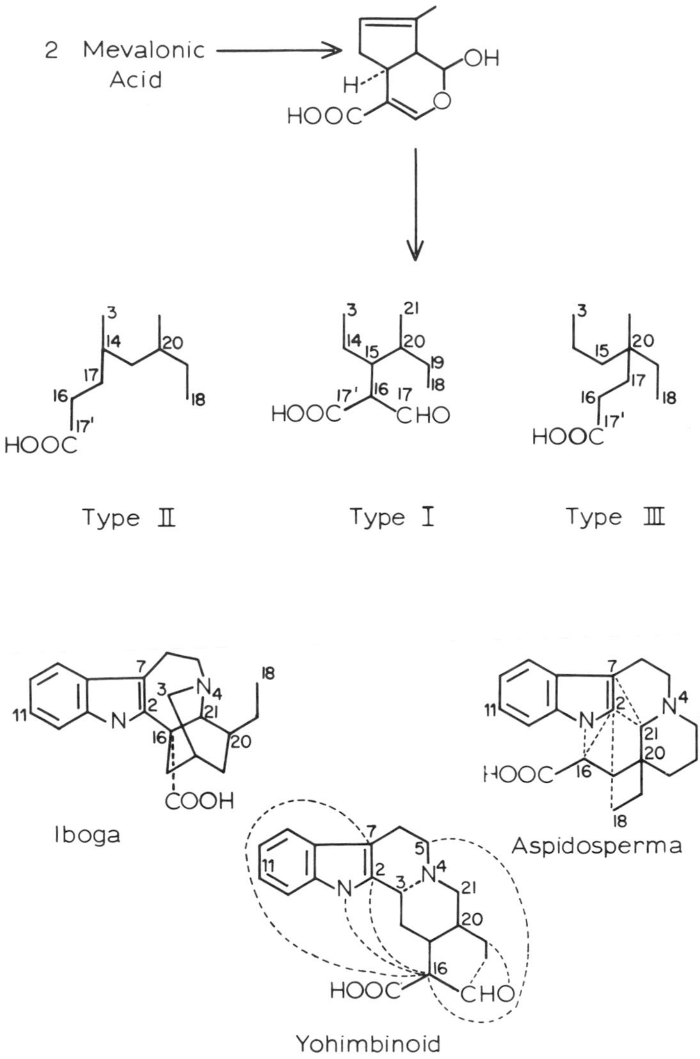

SCHEME 1 Types of complex indole alkaloids.

presumed equivalents in yohimbine, the presumption being
that this reflects their common genesis.[1,2] It should be noted
that positions 17 and 17' at some state could become equivalent
during biosynthesis. Also, the occurrence of racemic as well
as the optical antipodes in Type III alkaloids can be simply
explained if C-16 and C-17 are substituted for C-18 and C-19
by a rotation of C-20, i.e., in the absence of C-17'; there is
a possible equivalence of these two carbon units. This would
mean that the numbering system for optical antipodes would be
different; however, in the absence of definitive experiments,
no such distinction will be made at least for those compounds
lacking C-17'.

II. TYPE III BASES

A. The Vincamine Bases

The principal alkaloid of V. minor is vincamine III,
(Chart 2) whose structure became obvious when it was found
that treatment of the alkaloid with acid gave (-)-eburnamonine
V, the optical antipode of eburnamonine, a major base, from
Hunteria eburnea Pichon.[3] Dehydrogenation of (-)-eburna-
monine gave (-)-4-ethyl-4-propyl-4, 5-dihydrocanthinone, the
optical antipode of the (+)- form obtained from eburnamonine

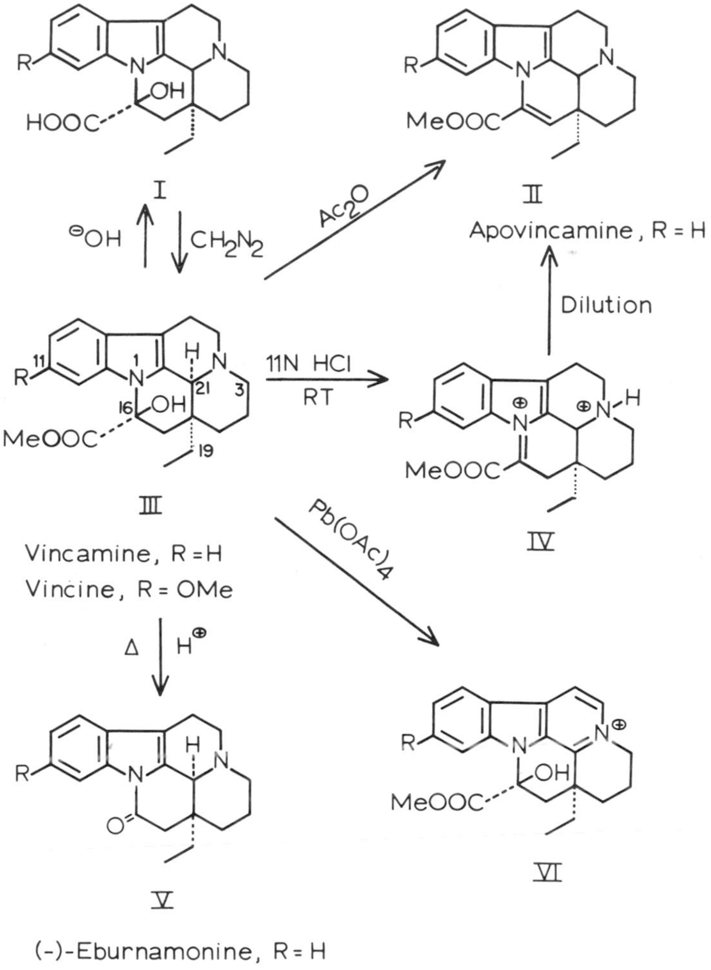

CHART 2 Some properties of vincamine and vincine.

whose structure proof is described below.[4]

11-Methoxyvincamine (vincine, III) whose ring A methoxyl is so placed because of its UV-spectrum and color reaction specific for that position, had properties which paralleled those of vincamine.[5] These overall structural conclusions were also reached independently upon the basis of mass

spectral comparisons.[6] Ring C quaternary tetradehydro

compounds VI were formed by lead tetraacetate oxidation and

digestion of VI (R=OMe) with potassium isoamylate generated

7-methoxy-β-carboline.[5]

The hydroxyl group of vincamine could not be acetylated;

instead, the apo compound (an N-vinylindole, II) was obtained.

In fact, simply heating the alkaloid in strong acid was all that

was necessary. This experiment finds an analogy in eburna-

mine (vide infra) and requires the intermediacy of the iminium

form (IV). This form can be detected in strong acid solution

via the UV-spectrum. Vincamine in 11 N hydrochloric acid

has a long wavelength maximum at 360 nm (log ϵ = 3.85) which

is lost upon dilution and formation of apovincamine.[6]

Lithium aluminum hydride reduction of vincamine yields

vincaminol.[7,8] The latter derivative upon reflux in 2 N hydro-

chloric acid is converted quantitatively into (-)-eburnamonine.[8]

Alternative procedures for obtaining (-)-eburnamonine are the

treatment of vincaminol with periodic acid and the oxidation of

vincaminic acid with ammoniacal silver nitrate.[4]

Although a cis fusion for the DE rings of vincamine (VII)

follows from the conversion to eburnamonine, independent

physical evidence has been sought and a total synthesis has

been realized. In the I.R.-spectrum in the 3.4 μ region, there

is a single band characterisic of a <u>cis</u>-fused quinolizidine

system (rings CD) and in the NMR-spectrum, the peak for the

C-3 proton is at 3.92 ppm,[10] a position characteristic of C-3

hydrogens in analogous <u>cis</u> fused quinolizidines such as 3-epi-

yohimbines.[11] The pseudo-first order rates of methiodide

formation have also been used as a proof of configuration.[10]

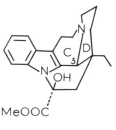

VII

Vincamine

16-Epivincamine has very similar IR and UV-spectra to

vincamine.[12] The mass spectrum with peaks at M, M-CH$_3$,

M-C$_2$H$_5$, M COOMe, M-H$_2$O, M 102, etc., were nearly

identical with those of vincamine differing somewhat in inten-

sity and having a peak at m/e 266 which in vincamine is found

at m/e 267. Reduction of the alkaloid gave 16-epivincaminol,

which after oxidation with periodic acid furnished (-)-eburna-

monine. Another simple proof for the C-16 isomerism was

obtained by dehydrating the alkaloid with methanolic hydrogen

chloride. The product was identical with apovincamine.

TABLE 2

Chemical Shifts of Some C-16 Substituted Dihydroeburnamenines

Compound	Substituents	Δ ppm
Vincamine	e-COOMe	3.82
Epivincamine	a-COOMe	3.70
Deoxyvincamine	e-COOMe a-H	3.89 4.68
Deoxyepivincamine	e-H a-COOMe	5.00 3.82
Isoeburnamine	e-H	6.00
Eburnamine	a-H	5.48

Having the epi-alkaloid made possible a useful NMR comparison of 16 substituted derivatives (Table 2) in which the downfield resonance of the protons of the quasi-equatorial substituents can be seen and for which structural assignments have been made.[13] Deoxyvincamine and deoxyepivincamine were produced by the hydrogenation of apovincamine in a ratio of about 1:9. Deoxyepivincamine after hydrolysis and re-esterification was converted into deoxyvincamine.[13]

The absolute configuration of vincamine and hence the other eburnamine alkaloids has been accomplished by showing that the optical rotatory dispersion curve of (-)-1,1-diethyl-1,2,3,4,6,7,12,12b-octahydroindolo [2,3-a] quinolizidine

VIII   IX   X

(VIII), mp 105-106° obtained by a Wolf-Kishner reaction on (-)

-eburnamonine was enantiomeric with (+)-1, 2, 3, 4-tetrahydro-

harman (IX).[14] The R-configuration has to be ascribed to the

latter compound because it has been converted into N-carboxy-

ethyl-D-alanine (X) of known absolute configuration.[14] This

work establishes the absolute configurations of the eburnamine

alkaloids since they have all been interrelated (vide infra).

Vincamine has been synthesized by a route which makes

its own contribution to the solution of a stereochemical pro-

blem and the results were in agreement with the above results

(Chart 3).[15]

Acid catalyzed condensation of tryptamine with the

aldehydic ester (XI) yielded a mixture of the tetracyclic lac-

tams (XII). Reduction of the lactam carbonyl was achieved

through conversion to the thiolactam esters with phosphorous

pentasulfide and subsequent desulfurization with Raney nickel

to give the amino esters XIIIa and b. The stereochemistry of

the amino esters could be decided on the basis of the ease of

oxidation of XIIIa by mercuric acetate; this was the isomer

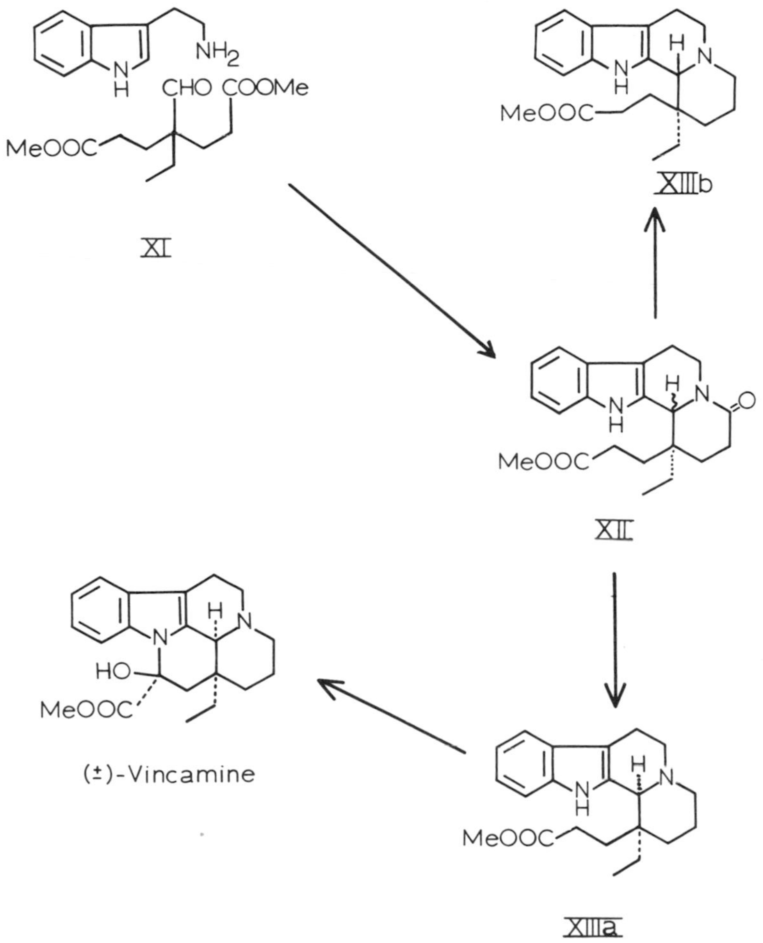

CHART 3 A synthesis of vincamine.

which was less rapidly eluted upon chromatography than XIIIb

and had $\nu_{c=o}$ 1735 cm^{-1} (XIIIb; $\nu_{c=o}$ 1725 cm^{-1}). Mercuric

acetate oxidation followed by reduction of the resultant

immonium salt with sodium borohydride, was a convenient

method for the interconversion of the amino esters.

Oxidation of the methylene group adjacent to the methoxy-

carbonyl group could be accomplished in low yield by treatment

of the amino ester XIIIa with a sodio-p-nitrosomethylaniline

and excess triphenylmethyl sodium then acid hydrolysis. This

procedure gave rise to (±)-vincamine in low yield. Oxidation

of isomer XIIIb furnished products which could be differenti-

ated from vincamine by thin layer chromatography.

Vincaminine and vincinine appeared to be on the basis of

their molecular formulas, group analyses and spectral data,

oxo analogs of vincamine and vincine, respectively. The C-19

position is assigned to the oxo group in vincaminine (XIV) on

XIV

Vincaminine, R = H

Vincinine , R = OMe

the basis of the formation of a hydrazone, formation of acetic

acid in the Kuhn-Roth oxidation, I.R. $\nu_{c=o}$ 1720 cm^{-1} and a

peak 3.61 τ characteristic of a methyl keton versus a triplet

(CH$_3$CH$_2$-) centered at 9.10 τ for vincamine. The overall

NMR pattern due to the aromatic protons is the same for vinca-

mine and vincaminine; the methyl protons of the ester group

appear at 6.22 τ in both alkaloids. All this points to a similar

topology for both alkaloids and the mass spectra are

essentially identical, if the mass of the carbonyl oxygen is

allowed for.[16]

A similar analysis settles the structure of 19-hydroxy

vincamine.[13] The stereochemistry of the 19-hydroxyl remains

to be established.

B. The Eburnamine Bases

Very recently, many years after the studies on the major

alkaloids of V. minor have been underway, eburnamine, iso-

eburnamine and eburnamonine have been isolated from a

strongly polar mixture of worked up bases. These bases are

enantiomeric in their ring configurations with their methoxy-

carbonyl congeners, vincamine, vincaminine, vincine and

vincinine. Perhaps (±)-eburnamonine (vincanorine) originates

as an accidental by-product of the oxidation of (iso)eburnamine

and oxidative demethoxycarbonyation of vincamine. The

amounts of the above eburnamine bases isolated were very

small and the identifications depend upon the comparison of

relevant physical data of which the optical rotations and mass

spectras were crucial.

The chemistry of the eburnamine group of bases,

especially eburnamonine, is the foundation upon which the

original vincamine structure is based and therefore is worth

summarizing. These bases were first recognized in <u>Hunteria</u>

<u>eburnea</u> Pichon.[18]

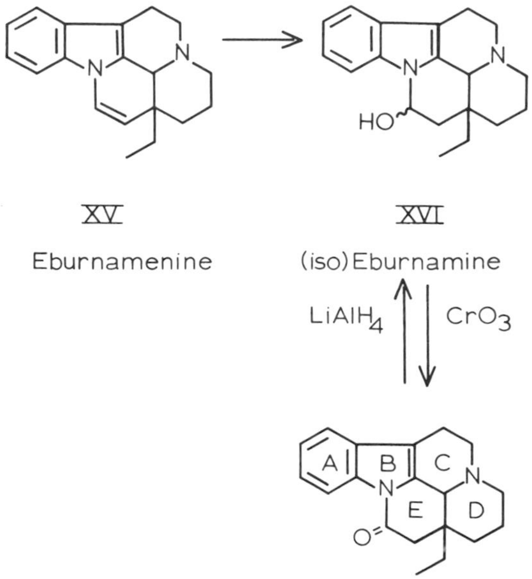

<u>XV</u>

Eburnamenine

<u>XVI</u>

(iso)Eburnamine

LiAlH$_4$ CrO$_3$

<u>XVII</u>

Eburnamonine

Eburnamine and isoeburnamine, are diastereoisomeric

pentacyclic indoles (XVI) convertible by acids into eburname-

nine, and N-vinylindole (XV), on the one hand, and by chromic

acid into eburnamonine, an N-acylindole (XVII), on the other.

Reduction of eburnamonine with lithium aluminum hydride

regenerated the alcohols, XVI.[19] When eburnamonine was

heated with selenium at 360° for 5 minutes, it gave an almost

quantitative yield of 4-ethyl-4-propyl-4, 5-dihydrocanthin-6-

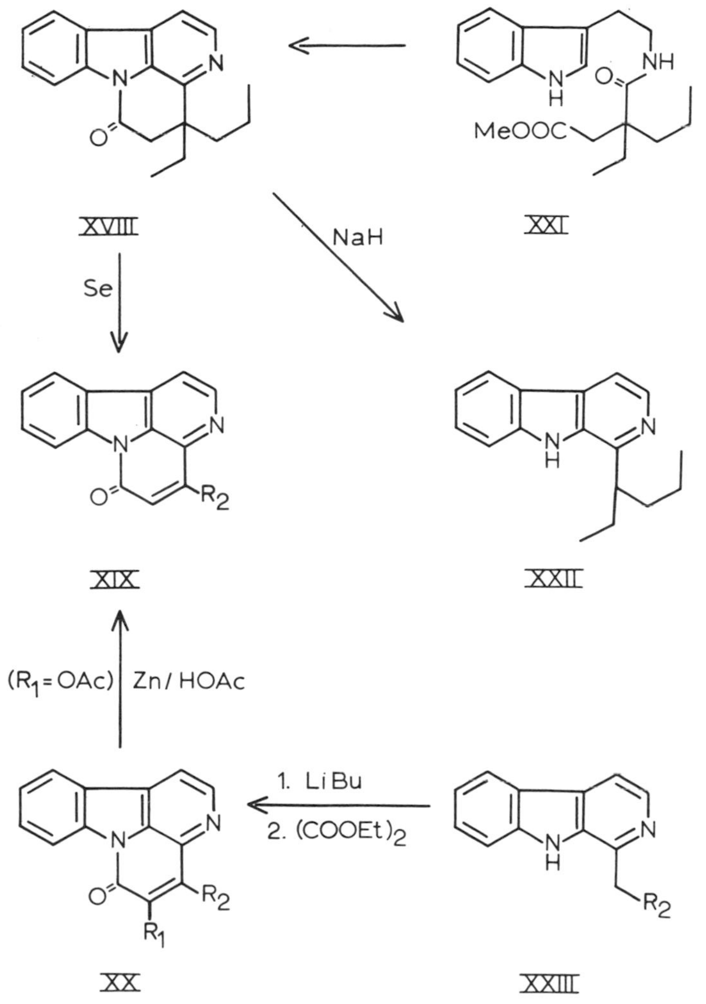

CHART 4 Synthesis and properties of selenium degradation
products of (+)-eburnamonine.

one (XVIII). Prolonged heating of XVIII with selenium eventu-

ally gave the two possible canthin-6-ones XIX (R_2=Et or Pr),

which were synthesized by condensation of diethyl oxalate and

the dilithium derivative of the appropriate 1-alkyl-β-carboline

(XXIII, R_2=Et or Pr) followed by reductive removal of the 5-

hydroxyl group in the product (XX, R_1=OH, R_2=Et or Pr).

When XVIII was subjected to prolonged reflux with sodium

hydride in toluene, it extruded the acetyl moiety to form (-)-

1-(1-ethylbutyl)-β-carboline (XXII), the racemic modification

of which was readily synthesized. It should be noted that the

optically active form cannot be prepared from the resolved

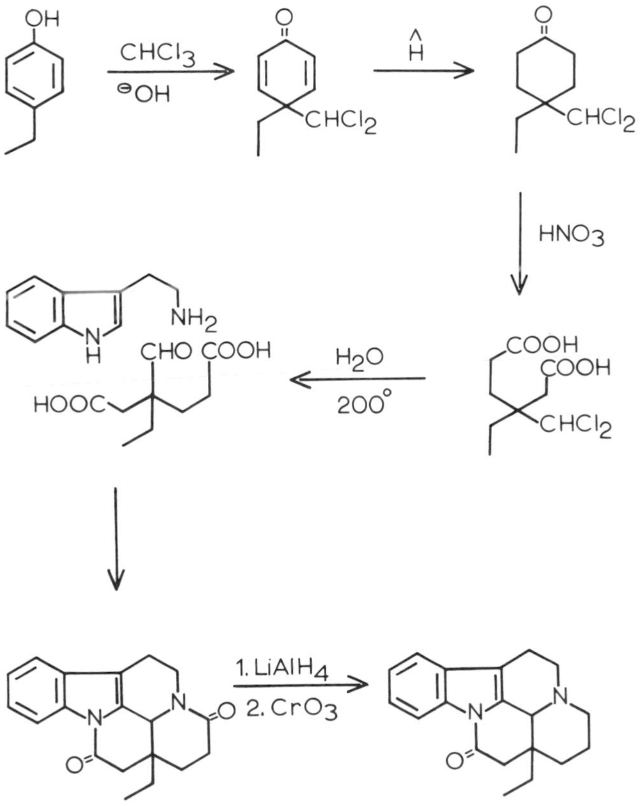

CHART 5   A total synthesis of (±)-eburnamonine.

amide under Bischler-Napieralski conditions. A synthesis of racemic XVIII has also been accomplished by ring closure of the amide, XXI.

These results did not unequivocally extablish the structure of eburnamonine. This was secured by a degradation of eburnamine (vide infra) as well as by a total synthesis (Chart 5).[19,20] Condensation of β-ethyl-β-formyladipic acid with tryptamine gave in one step (±)-eburnamonine lactam which upon reduction with lithium aluminum hydride followed by oxidation of the resultant (±)-eburnamines (XVI), afforded (±)-eburnamonine. The desired aldehydodicarboxylic acid was prepared in four steps from p-ethylphenol, as outlined in Chart 5. (±)-Eburnamonine (vincanorine) has been resolved into its antipodes by use of dibenzoyltartaric acid.[16]

A second synthesis has been realized via the condensation of ethyl bromoacetate with XXIV (Chart 6). The immonium salt XXV in buffered solution gave the lactam XXVI, reduction of which by either chemical or hydrogenative means would be predicted to lead stereoselectively to the trans system XXVIII since the angular ethyl group in this nearly planar substance would be expected to have a strong effect. Hydrogenation, as well as sodium borohydride reduction of XXVI gave a single product, (±)-epieburnamonine, XXVIII. Hydrogenation or

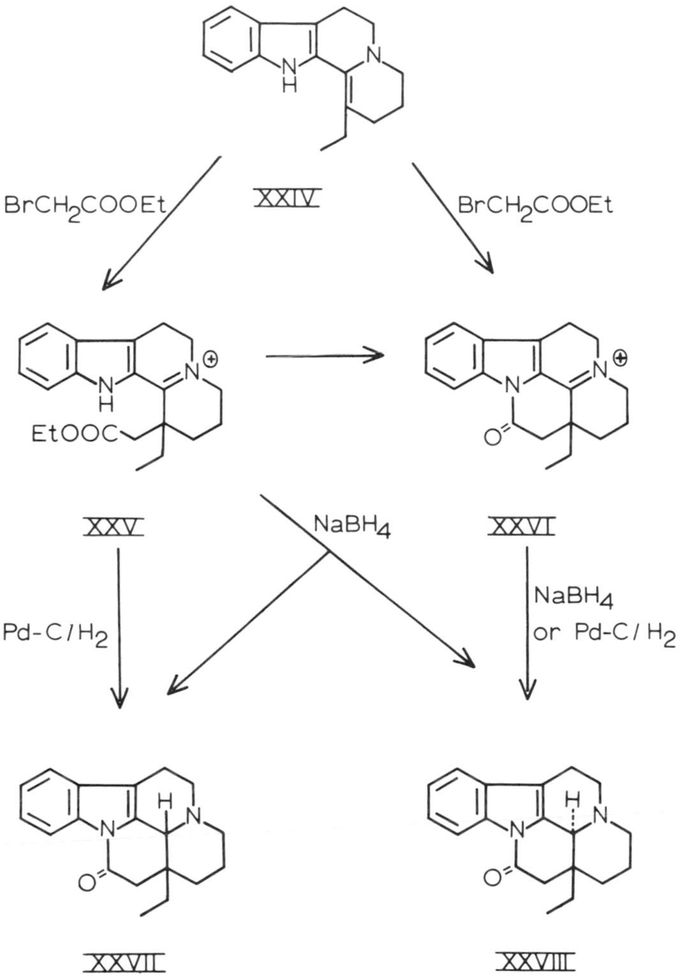

CHART 6   A second synthesis of (±)-eburnamonine.

sodium borohydride reduction of XXV yielded a mixture of

eburnamoninic and epieburnamoninic esters, alkaline treat-

ment of which led to (±)-eburnamonine (XXVII) and (±)-epi-

eburnamonine (XXVIII). While sodium borohydride afforded

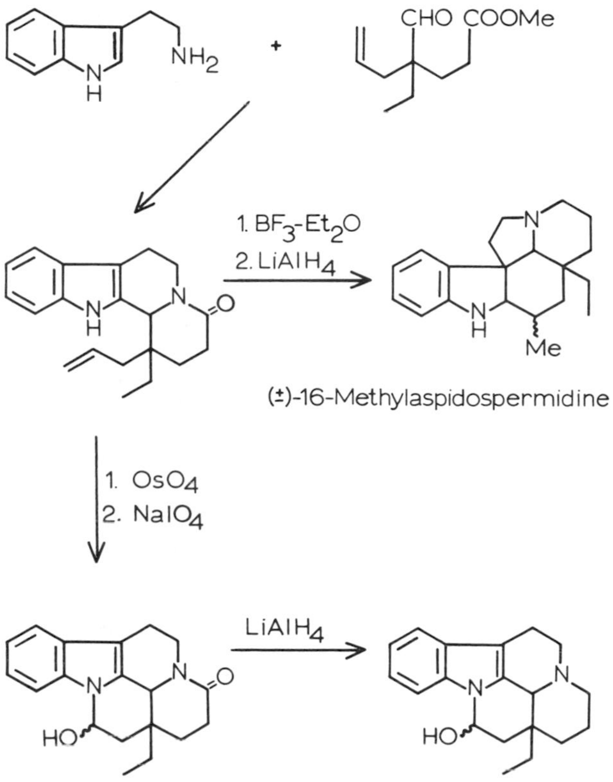

CHART 7 A route to eburnamine and 16-methyl aspidospermidine from a common intermediate.

a 1:1 mixture of products, hydrogenation yielded predominant-

ly (±)-eburnamonine.

The ability to rearrange substituted tetrahydro-β-carbolines

into indolenines under appropriate acidic conditions has led to

a synthesis of 16-methyl-aspidospermidine from an intermedi-

ate in another total synthesis of the eburnamine type alkaloids

(Chart 7).[23] This route to (±)-eburnamine is another variant

of the syntheses sketched in Charts 4 and 5. The boron tri-
fluoride-etherate rearrangement and ring closure of the tetracyclic intermediate appeared to have given an entirely homogeneous product and the subsequent lithium aluminum hydride
reduction was also stereospecific. These results compare
well with the results obtained in the total synthesis of aspidospermine where the stereospecificity, the reasons for it, and
the consequences were noted.[24]

Eburnamonine can be hydrolyzed to an amino acid which
recyclises with great ease. This tendency to cyclise is such
that lithium aluminum hydride reduction of methyl eburnamon-
inate gave only the ring closed alcohols, eburnamine and isoeburnamine.[19]

Wolff-Kishner reduction of eburnamine yields (+)-1,1-
diethyl-1,2,3,4,6,7,12,12b-octahydroindolo [2,3-a] quinolizidine (XXIX; R=H$_2$). This last compound was catalytically
dehydrogenated and rereduced to furnish the racemic compound.
It was synthesized by condensing tryptamine with 4-ethyl-4-
formyl-hexanoic acid and by reducing the resulting lactam
(XXIX; R=O) with lithium aluminum hydride.

Neither eburnamine nor isoeburnamine showed any properties, with the exception of the Wolff-Kishner reaction, which
would indicate their being in equilibrium with the theoretically

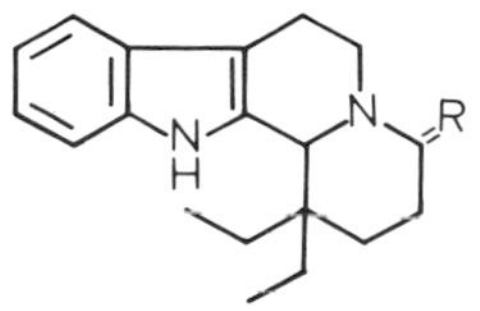

XXIX

R = H$_2$ or O

tautomeric aldehyde;however, the reactions to be described

imply the intermediacy of another tautomer, the iminium ion.

The reactions are summarized in Chart 8 (partial formulas).

If either eburnamine or isoeburnamine was allowed to stand at

room temperature in 0.5 N sulfuric acid for a period, it gave

rise to a mixture of about 90% of the former and 10% of the

latter and, if the solution was warmed briefly, eburnamenine

was obtained. The driving force in this reaction probably lies

in the relief of strain (1,3-diaxial interactions) in going from

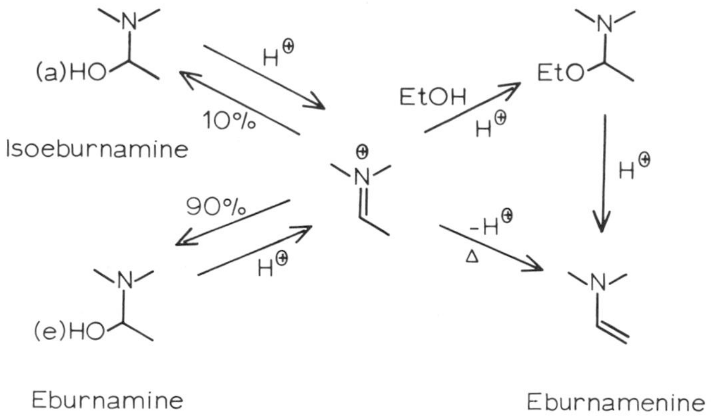

CHART 8 Interrelationships between eburnamine, isoeburnamine and eburnamenine (partial formulae).

isoeburnamine to eburnamine. In ethanolic picric acid, both alcohols gave one and the same O-ethyleburnamine (Chart 8), which, if the solution was refluxed for a short period or the picrate itself was heated above its melting point, furnished eburnamenine in quantitative yield. Water was also eliminated from (iso)eburnamine methiodide(s) when it was crystallized from water.[19]

In agreement with these results, the alcohols were un-affected by sodium borohydride and only very slowly reduced to dihydroeburnamenine by lithium aluminum hydride because the reaction conditions are poor for the formation of the iminium ion. It is for this reason that dihydroeburnamenine was best prepared by catalytic reduction of the olefin.

Eburnamenine reacted readily with osmium tetroxide to form an amorphous glycol that, upon oxidation with chromic oxide, gave hydroxyeburnamonine, which resisted further attack by the oxidant.

In view of the foregoing described acid-catalyzed trans-formations and the fact that inorganic acids were used in the isolation of the alkaloids from plant material, it is quite possible that the ratios of isoeburnamine to eburnamine and eburnamenine might be different in situ even to the extent of the exclusion of the last two so-called alkaloids.

C. Mass Spectra of the Eburnamine-Vincamine Alkaloids

Upon volatilization into the mass spectrometer, three alkaloids--eburnamenine, eburnamine, and isoeburnamine-- gave the same spectrum because of the facile loss of water from the last two alkaloids.[6, 25] Apovincamine gave results that were equivalent to those of eburnamenine, if allowance was made for the extra 59 mass units.[6] Eburnamine, for example (Chart 9), showed a weak peak loss of the C-3 proton [cf. yohimbine class],[26] and the major primary fission was a retro Diels-Adler-like cleavage of ring C, then loss of one of the two chains ("a" or "b" split) to yield the major peaks m/e 208 and 249. Eburnamonine split likewise, and in addition exhibited a further peak m/e 237 which was the mass of the fragment M-29 from which carbon monoxide had been ejected. These conclusions were confirmed by running the mass spectra of suitably labeled compounds prepared as follows: with lithium aluminum deuteride, eburnamine gave dihydroeburnamenin-16-d, and eburnamonine gave dihydroeburnamenin-16-d_2, and the oxygen of eburnamonine was exchanged for ^{18}O.[25]

Dihydroeburnamenine behaved analogously to eburnamonine, and its two most prominent peaks apart from M and M-1 were M-29 (loss of ethyl) and M-70 ("a" split).[25]

Vincamine showed peaks at M, M-1, M-CH$_3$, M-C$_2$H$_5$,

M-COOMe, M-H$_2$O, and M-HCOOCH$_3$. Since the last two

peaks were equivalent to apovincamine and eburnamonine, it

was not surprising to observe their characteristic fragmenta-

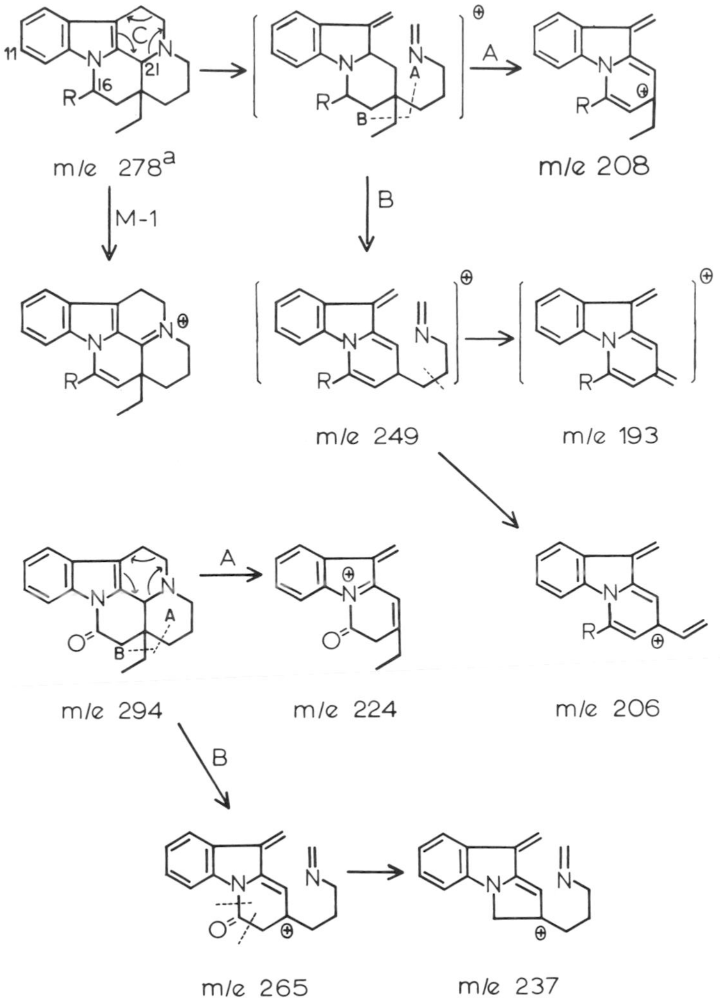

CHART 9 Principal electron impact products of eburnamenine,
apovincamine and eburnamonine.

[a]The m/e is given for R=H; for R=COOMe, add 59 mass units.

tion peaks in the line-rich spectrum. 11-Methoxy and 19-oxo congeners behaved similarly to vincamine except the expected shifts in the peaks due to +30 and +14 mass differences.[6]

D. The Quebrachamine Group

The elucidation of the structures of numbers of this group was facilitated greatly by the knowledge of the structure of quebrachamine to which they can all be related. Since quebrachamine is a 7,21-seco-aspidospermidine, the close relationship between representatives of these two classes in the alkaloids of V. minor is not surprising. The Vinca bases have stereochemistry equivalent either to (+)-quebrachamine or are racemic.

The first fixing of the molecular architecture of (-)-quebrachamine (XXX) was a consequence of the structure determination of aspidospermine.[27, 28, 29] In work prior to this, it had been shown that the two alkaloids were related, although one

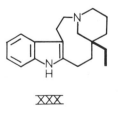

XXX

(-)-Quebrachamine

XXXI

a. R = OH
b. R = OOH
c. R = CN

was an indole and the other an indoline, since both gave upon

zinc dust distillation a mixture from which 3, 5-diethyl- and 3-

ethyl-5-methylpyridine could be isolated as the mixed picrate.

[28, 30] Oxidation of quebrachamine under a variety of conditions

gives products which are probably 7-substituted-indolenines.

For example, ozone and peracids give a hydroxy base probably

XXXIa; catalytic oxidation gives XXXIb which is readily con-

verted into XXXIa. [29, 30] Yet another indolenine XXXIc was

obtained by the action of cyanogen bromide. [31] Lithium aluminum

hydride reduction of XXXIa or hydrolysis of XXXIc gave back

quebrachamine.

It was an early application of mass spectrometry to alka-

loid chemistry that first established the structure of quebrach-

amine. [32] Initially the zinc distillation products were separated

by vapor phase chromatography and were found with the help

of mass spectrometry, to be 3-ethylpyridine, (75%), 3-methyl-

5-ethylpyridine (12%), 3-ethyl-4-methylpyridine (5) and 3, 5-

diethylpyridine (5%) along with 3-methylindole, 2-ethylindole,

2, 3-dimethylindole and methyl ethylindole. The major frag-

ment together with 2, 3-diethylindole account for all the carbons

and both nitrogens of quebrachamine. In order to provide a

model with identical ring structure, aspidospermine was con-

verted into 12-methoxyquebrachamine and 12-methoxymonodeu-

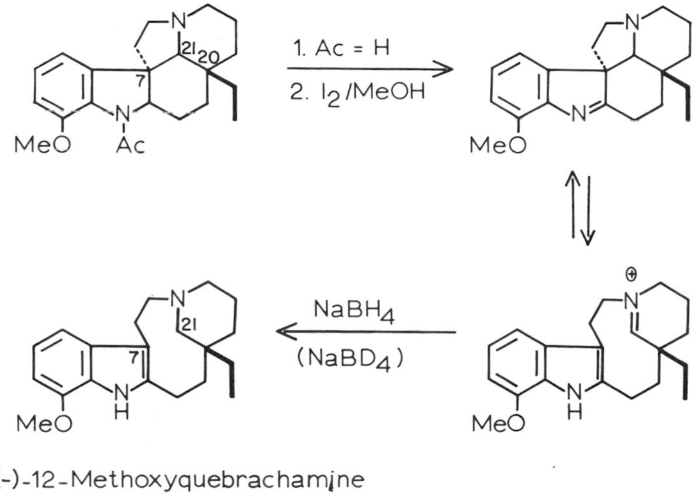

(-)-12-Methoxyquebrachamine

CHART 10 Conversion of aspidospermine to 12-methoxy-
quebrachamine.

terioquebrachamine as shown in Chart 10. Comparison of the

mass spectra of 12-methoxyquebrachamine and quebrachamine

left no doubt as to the identity of the hydroaromatic portions of

the molecules (Chart 11). Fragments containing the aromatic

nucleus were shifted up +30 mass units in the methoxy deriva-

tive with respect to the same products from the unsubstituted

compound. In the monodeuterio compound the deuterium show-

ed up in a piperidinic peak (C-21 deuterium) as expected.

12-Methoxyquebrachamine has a rotation of -103° in diox-

ane compared with -111° found for quebrachamine in the same

solvent, and the optical rotatory dispersion curves are very

similar. It was deduced that (-)-quebrachamine has the same

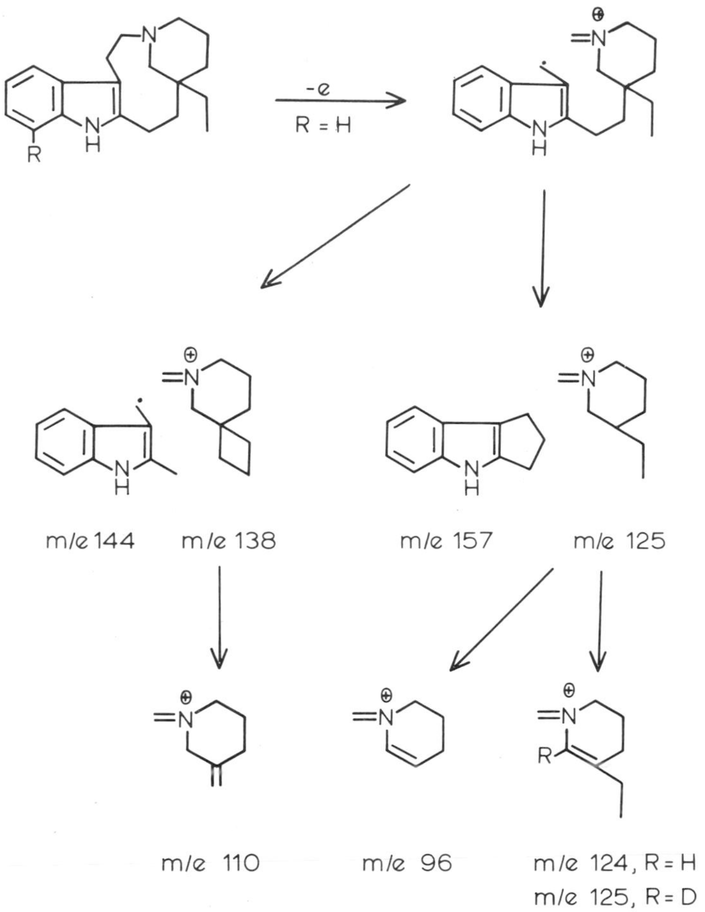

CHART 11 Important breakdown products of quebrachamine and
         12-methoxyquebrachamine upon electron impact.

absolute configuration at position 20 as does aspidospermine.

A number of syntheses of quebrachamine (and aspidospermine) are now available. The synthesis sketched in Chart 12
was the first and is readily adaptable to the preparation of
other analogs. Other synthetic approaches have been touched

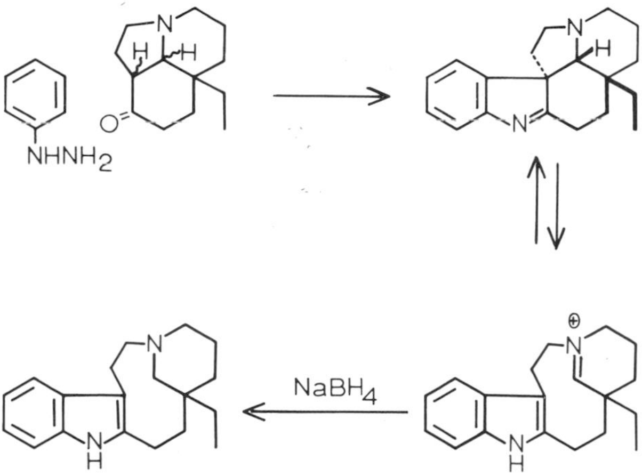

CHART 12   A synthesis of (±)-quebrachamine.

upon indirectly in this chapter, see Charts 7 and 14.

The absolute stereochemistry of this group of compounds

follows from extensive optical rotatory dispersion studies in

which quebrachamine-aspidosperma bases were compared with

strychnoid compounds of unknown absolute configuration.[33]

As far as possible, all structures in this chapter are drawn to

show these stereochemical points.

The (+)-quebrachamine congeners of V. minor have been

interrelated as shown in Chart 13. The chemistry has con-

cerned itself with three types of reactions; firstly, methylation

of quebrachamine and vincadine to the equivalent N-methylin-

doles; secondly, hydrolysis and decarboxylation of vincadine,

vincaminorine, and vincaminoreine in refluxing 5 N hydrochlor-

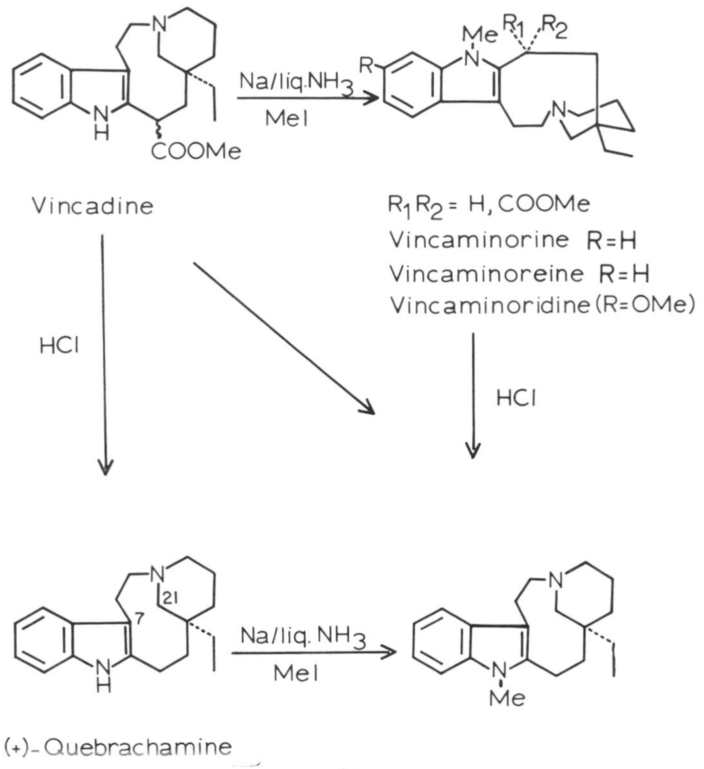

Vincadine

R₁R₂ = H, COOMe
Vincaminorine R=H
Vincaminoreine R=H
Vincaminoridine (R=OMe)

(+)-Quebrachamine

CHART 13 Interconversions of quebrachamine, vincadine,
vincaminorine and vincaminoreine.

ic to their respective demethoxycarbonyl bases; and thirdly,

base catalyzed epimerization of vincaminoreine and vincaminor-

ine into an equilibrium mixture in which vincaminoreine pre-

dominates. Vincaminoridine is believed to be 11-methoxyvinca-

minoreine and has been synthesized (Chart 14) but a comparison
                                                        36
with the natural product could not be made.

Pseudo first order rates of methiodide formation with

vincaminorine (pK'$_a$ 5. 45) and its C-16 epimer (pK'$_a$ 7. 55) show

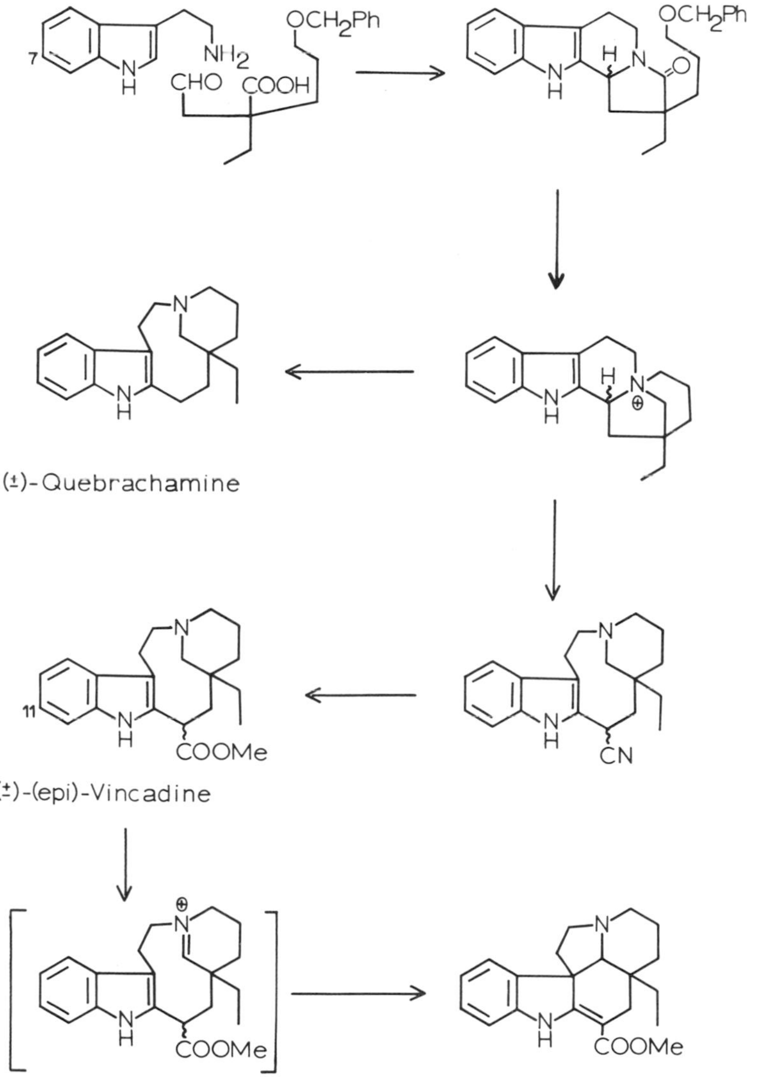

CHART 14 Synthesis of (±)-vincadine and (±)-vincadifformine
(vincaminoridine, 11-methoxy-16-epivincadine,
was prepared from 7-methoxytryptamine).

that in both cases the basic nitrogen is hindered. Of particular

interest is the opinion that a peak at 6.25 ppm in the NMR

spectrum of vincaminorine representing one proton is due to

the C-16 hydrogen being strongly deshielded by N-4. In

vincaminoreine the same proton has its signal absorbing at

3.9 ppm. [37] From this and mass spectrometric data differing

deductions have been made as to the configurations of the C-16

epimers. [35] These arguments depend in the end upon assump-

tions concerning the stable conformation of the azabicyclodo-

decene system and remain to be resolved.

A new approach to the total synthesis of (±)-quebrachamine

has also given rise to (±)-vincadine and (±)-16-epivincadine

(Chart 14), and therefore in a formal sense (see Chart 13)

vincaminorine and vincaminoreine. The synthetic (±)-vinca-

dine was also converted into (±)-vincadifformine and minovine

(Chart 14). [36]

E.   The Vincadifformine Group

Alkaloids of this group are characterized by their β-ani-

linoacrylate-like UV spectra and high negative rotations. The

compounds (see Table 3) were interrelated and their structures

determined via an analysis of combustion data, UV, IR, NMR

spectra and comparison of their mass spectra with (-)-14,15-

TABLE 3

Relationships Between Some Aspidosperma Bases
Occurring in Vinca Species

	$[a]_D$	R_1	R_2	R_3	Other
Vincadifformine	- ±	H	H	H_2	
Minovine	±	H	Me	H_2	
Minovincine, Vincorine	-	MeO	H	H_2	
Minovincinine	-	H	H	H, OH	
Minoricein	-	MeO	H	O	
Tabersonine	-	H	H	H_2	Δ 14, 15
11-Methoxytabersonine	-	MeO	H	H_2	Δ14, 15
Lochnerinine	-	MeO	H	H_2	14, 15-epoxy

dihydrotabersonine which was found to be enantiomeric with

(±)-vincadifformine, (see Chart 15).

(-)-Tabersonine upon hydrogenation gave (-)-14, 15-dihy-

drotabersonine [(-)-vincadifformine] which upon heating at 110°

with 4N hydrochloric acid yielded (-)-demethoxycarbonyl-14,

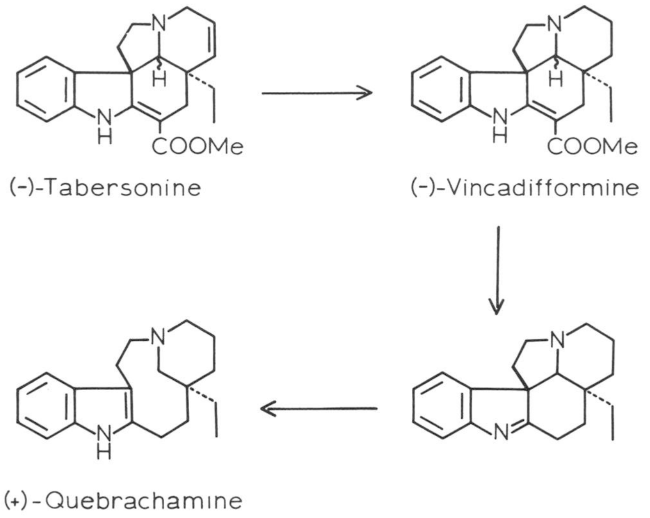

(-)-Tabersonine   (-)-Vincadifformine

(+)-Quebrachamine

CHART 15   Degradation of (-)-tabersonine to
(+)-quebrachamine.

15-dihydrotabersonine [antipode of (+)-1, 2-dehydroaspidosperm-
idine].[40] Reduction of the dehydro compound with potassium

borohydride furnished (+)-quebrachamine. Alternatively (+)-

quebrachamine could be obtained by conversion of (-)-taber-

sonine to demethoxycarbonyltabersonine followed by potassium

borohydride reduction and catalytic hydrogenation of the inter-

mediate 14, 15-dehydro-quebrachamine.[38]

(±)-Vincadifformine has also been converted to (±)-de-

methoxycarbonylvincadifformine and hence to (±)-quebracha-

mine which was found to be identical in all respects with (±)-

quebrachamine prepared by mixing the naturally occurring

antipodes.[39]

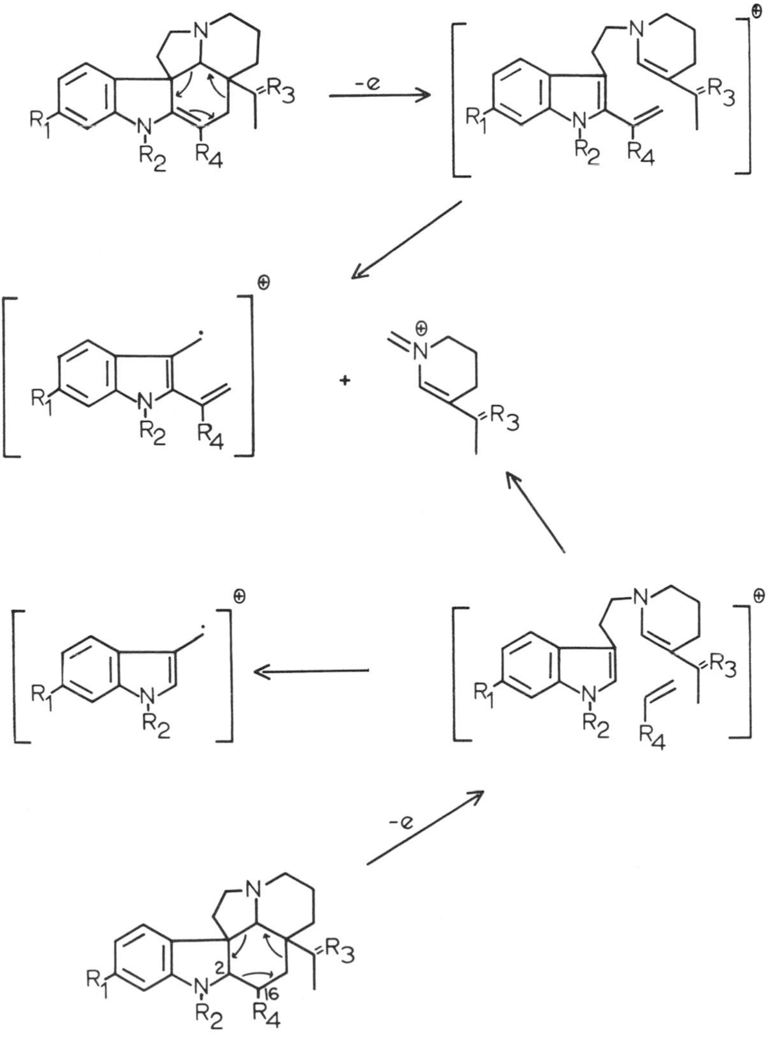

CHART 16 Major fragments produced upon electron impact
for vincadifformine and derivatives (R_1=H, OMe;
R_2=H, Me; R_3=H_2; HOH, O; R_4=COOMe, CH_2OH).

A further support for the vincadifformine structure came

from an examination of the mass spectrum of 2,16 dihydro-

vincadifformine (Chart 16).[41] These compounds fragment as expected for a substituted aspidospermine. The 2, 16-dehydro-compound was prepared by a zinc-sulfuric acid reduction of vincadifformine.

From V. erecta Russian workers have obtained a base which has been named ervamine.[42] Its properties would make it appear to be (-)-vincadifformine; however, its stated conversion to (-)-quebrachamine would suggest that the stereochemistry at C-20 must be antipodal. The writer confesses to some difficulties in interpreting the Russian work on the alkaloids of V. erecta.

The absolute and relative configurations dipicted for vincadifformine and its congeners follows from relating the molecular rotations and rotatory dispersions with analogous bases in the strychnine group.[33] The configuration at C-21 has not yet been determined.

F. Kopsinine and Pseudokopsinine

From the physical data and combustion figures, pseudokopsinine, another base isolated from aerial parts of V. erecta,

Kopsinine

could very well be the 16-epimer of kopsinine which is also obtained from the same plant.[73]

III. TYPE I BASES

In Chart 17 are sketched the skeleta of the various classes of Type I bases distributed among <u>Vinca</u> alkaloids. The chemistry of some of these bases such as reserpine and reserpinine (pubescine), akuammicine and congeners (vinervine, vincanine, vincanidine), akuammine (vincamajoridine) and relatives [strictamine (vincamidine) and vincoridine] will either not be dealt with or only touched on lightly. Their structural elucidation was independent of <u>Vinca</u> alkaloids as a whole, and their chemistry has not been crucial for the study of other <u>Vinca</u> bases.

Several representatives of the ajmaline-sarpagine class, namely, the 16-carbomethoxytetraphyllicines, have been useful in expanding our knowledge of this alkaloid subclass (see Chart 19). The examples of herbaceine, hervine and herbaline widen our knowledge of plant biosynthetic capability.

A. The 16-Methoxycarbonyltetraphyllicine Group

Detailed knowledge of the ajmaline-sarpagine subgroup is intimately connected with the extensive investigations of

Yohimbine

Ring-E-oxygen heterocycle

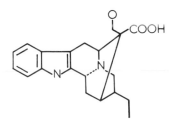

17,18-seco-Yohimbine

Akuammine

Sarpagine

Ajmaline

Chart 17   Type I alkaloid skeleta found in <u>Vinca</u> alkaloids.

Rauvolfia Bases.[43] Most of these bases have been interrelated

and their detailed structures rest on a number of separately

derived proofs.   Two of these were X-ray crystallographic

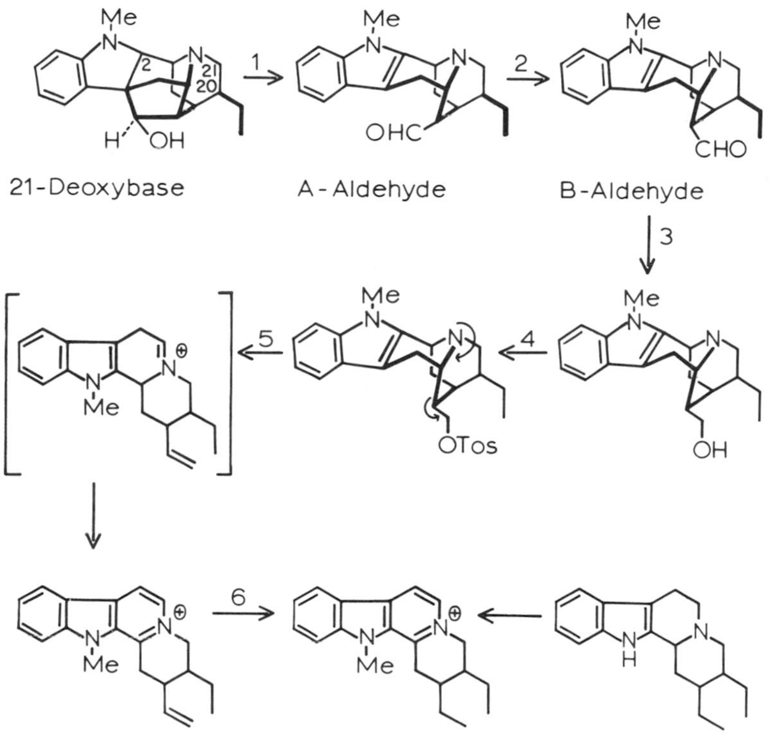

21-Deoxybase   A-Aldehyde   B-Aldehyde

Reagents: 1. Pb(OAc)$_4$; 2. Base; 3. NaBH$_4$; 4. Tosyl
Chloride/Py; 5. Collidine; 6. P$_2$/Pd.

CHART 18 Degradation of deoxyajmaline and deoxyiso-
ajmaline (epimeric at C-20).

analyses of sarpagine derivatives and a third was a facile un-

folding of the ajmaline molecule by stereospecific processes

(Chart 18) which led to a ring E-seco compound of known ab-

solute configuration,[44] viz. 1-methyltetradehydrocorynantheane

from 21-deoxyisoajmaline, whose absolute stereochemistry

had been established via a correlation of dihydrocorynantheol with cinchonamine and quinine.[45] A synthesis of ajmaline has

also been realized.[46]

The orientation at C-2 was solved by preparing an indol-

enine which upon catalytic reduction furnished a C-2 epimer

as judged by its optical rotatory dispersion curve which was

in a mirror image relationship to that of 21-deoxy-ajmaline.

Comparison of the dissociation constants of the epimeric pair

showed that XXXII where the proton in the conjugate acid could

hydrogen bond to N-1 was a stronger base, pK'_a 8.44 than

XXXIII pK'_a 7.80 where this was not possible.

XXXII   Indolenine   XXXIII
21-Deoxyajmaline-   2-epi-21-Deoxyajmaline-
17-O-acetate   17-O-acetate

The structures of 16-methoxycarbonyl-2-epitetraphyllicine,

its 17-O-acetate and 1-demethyl analog have been secured by

conversion into 10-deoxysarpagine derivatives, with valuable

support from mass spectrometry using significant model com-

pounds for comparison purposes. The important interrelation-

ships are sketched, Chart 19, the compounds of reference

structure being akuammidine and polyneuridine methiodides

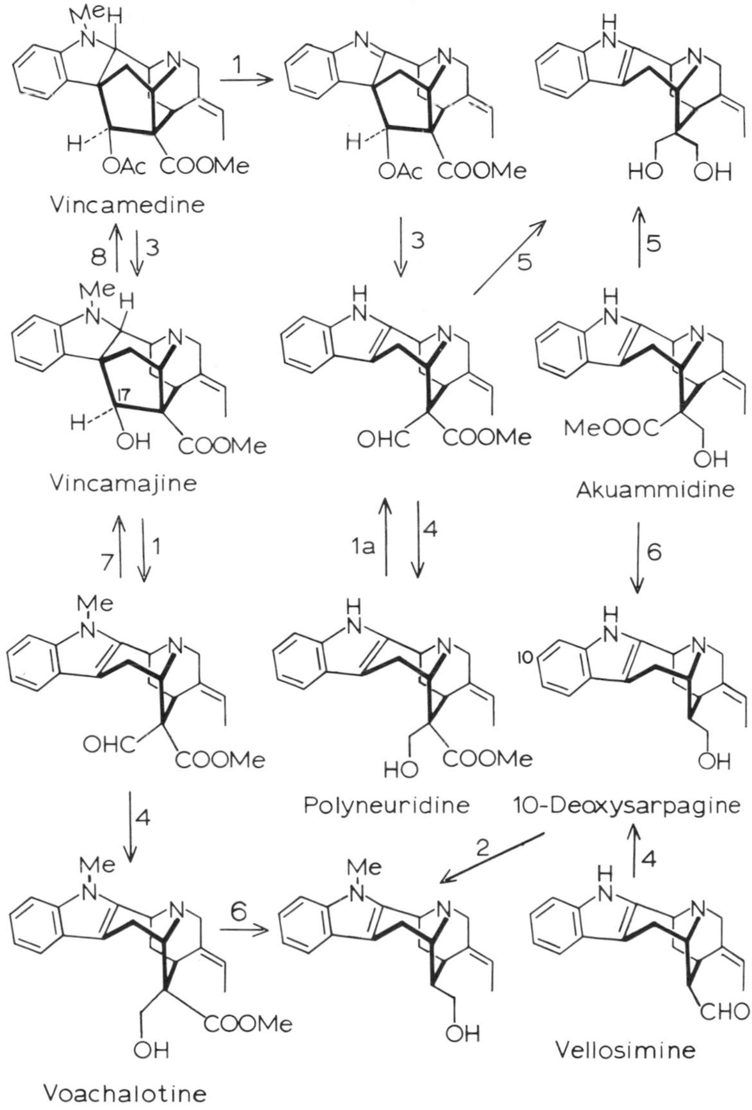

Reagents: 1. CrO_3/Py; 1a. CrO_3/acetone; 2. Na/NH_3/MeI;
3. Base; 4. KBH_4; 5. $LiAlH_4$; 6. Oppenauer, then
$LiAlH_4$; 7. Pt/H_2/HOAc; 8. Ac_2O.

CHART 19 Some interrelationships between the 16-methoxy-
carbonyl bases and 10 deoxysarpagine.

(X-Ray diffraction analysis)[47] and the 10-deoxysarpagines (via

correlation with ajmaline).[44]

The reconstitution of vincamajine by catalytic hydrogena-

tion of its corresponding indole aldehyde (Chart 19) follows

from the fact that 21-deoxyajmalal-A under similar conditions

furnished 2-epi-21-deoxy-ajmaline, i.e., the 17-hydroxyl

takes up the thermodynamically more stable of the two possi-

bilities.[48] A second proof for the configuration of the 17-

hydroxyl in these alkaloids came out of quebrachidine (1-de-

methylvincamajine) chemistry. Its lithium aluminum hydride

reduction product (a diol) gave in a facile manner an isopro-

pylidene derivative XXXIV which would not have been expected

from its 17 epimer.[49]

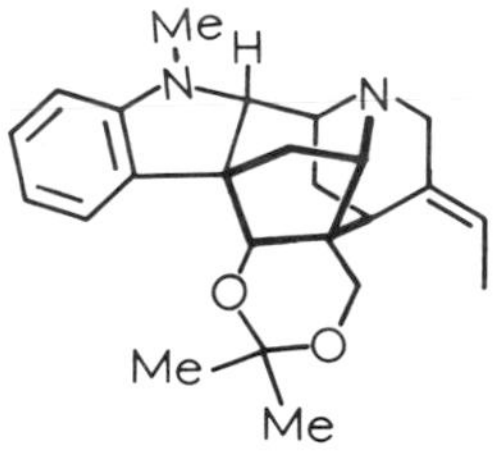

XXXIV

Vincarine is stated[50] to have a formula and mass spectrum

identical with quebrachidine [2-epi-17-methoxycarbonyl-1-

demethyltetraphyllicine]. It does belong to the ajmaline class

because lead tetra-acetate oxidation furnished an aldehyde

which after reduction yielded polyneuridine. This experiment

established unequivocally the complete stereochemistry of

vincarine with the exception of C-2 and C-17. If the mass

spectrum of vincarine is really closely similar to that of

quebrachidine as stated above, then the stereochemistry at C-2

has been established because C-2 epimers of the ajmaline

system have unique spectra clearly distinguishable from all

the other isomers (C-17, C-20, C-21) which are very much

alike.[51] Vincarine must therefore be quebrachidine or its

C-17 epimer. This, however, was not the conclusion of the

Russian authors who from some infrared data suggested that

vincarine might be the C-16 epimer of quebrachidine.

B. Tetraphyllicine Derivatives

It is interesting that vincamajoreine which is considered

to be 10-methoxytetraphyllicine[52] and its O-acetate are also

constituents of V. major and V. difformis. This sort of

isomerism has not yet been noted among the ajmaline-type

bases of the Rauvolfia species. The occurrence of such C-2

epimers in the same plant supports the idea that an indolenine

may be an intermediate in the biosynthesis. Majoridine was

shown to be 10-methoxytetraphyllicine-17-O-acetate from a

combination of UV, IR, NMR and mass spectrometric measurements.[53] A close overall correspondence (mass spectrometric shift technique) was observed between the following pairs: majoridine-tetraphyllicine-17-O-acetate, deacetylmajoridine-tetraphyllicine, and dihydrodeacetylmajoridine-21-deoxyajmaline. This parallelism, especially the two strong peaks at m/e 182 and 183 which are characteristic of ajmaline systems with a normal C-2 stereochemistry, i.e., the C-2 hydrogen cis to the C-17 - C-17 bond[51] eliminates the 2-epi configurational possibility which was up until this work, a characteristic of ajmaline-type bases from Vinca species. This was confirmed by comparing the optical rotatory dispersion curves for majoridine and tetraphyllicine, both bases showed a positive Cotton effect at about 260 nm. If the stereochemistry at C-2 had been different from tetraphyllicine, the sign of the Cotton effect would have been negative.[44]

The chemical shifts for the 17-O-acetate methyl and the C-17 hydrogen and its coupling to the C-16 hydrogen were analogous with those recorded for 21-deoxyajmaline 17-O-acetate.[44]

In view of the similarity between their physical data, deacetylmajoridine is believed to be identical with vincamajoreine.[53] In this connection, the recognition of 10-methoxy vellosimine (alkaloid Y) in V. major alkaloids fits very well biogenetically. This alkaloid's structure was fixed by two

step conversion into O-methyl sarpagine. Vellosimine has

been identified in V. difformis.[54]

C. Vincadiffine

Another base from V. difformis[54] has been shown to be a

2-acyl indole related to a growing number of compound of this

type which can be regarded as 3, 4-secosarpagines. Vinca-

diffine was obtained in very small quantity and has been as-

signed the structure 3-oxo-3, 4-secoakuammidine because of

its NMR and mass spectral characteristics. The configuration

of the C-16 substituents followed directly from the NMR since

the chemical shift 2. 57 ppm of the ester methyl could be inter-

preted as strong shielding by the aromatic group (Chart 20).

D. Mass Spectrometry of Ajmaline Derivatives

The mass spectra of ajmaline derivatives fall into three

types, namely, those that are similar to ajmaline, those that

possess a C-17 carbonyl function and those differing in the

stereochemistry at C-2.[51] This fundamental exploration into

the fragmentation properties of epimeric complex systems was

greatly facilitated by use of a high resolution mass spectro-

meter. Although the spectra are characteristic, they involve

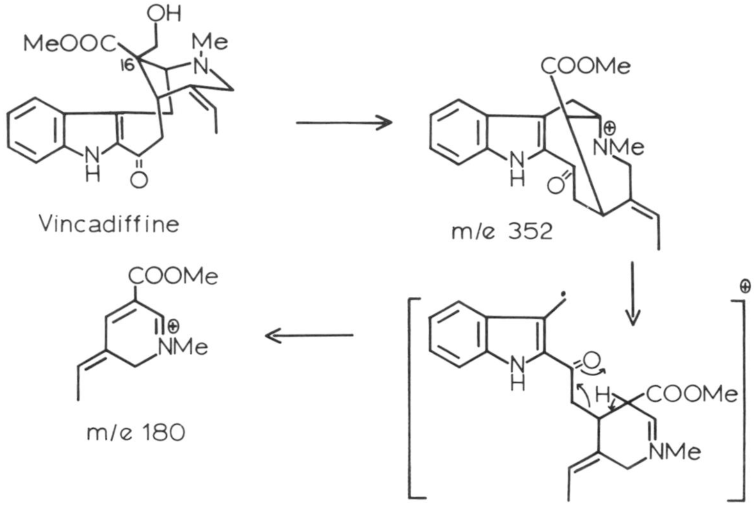

CHART 20 Principal fragmentation path for vincadiffine.

deep seated fissions since this compact highly branched
system in itself cannot support stabilized changes without
decomposition.

For example, quebrachidine split to yield strong peaks at
M-130, 130 (indoline + 1 carbon atom) and M-130-CH$_3$OH.
Vincamajine 17-O-acetate split similarly to furnish the observed strong peaks at M-144, 144(N-methyl indoline +1 carbon atom) and M-144-CH$_3$ OH.

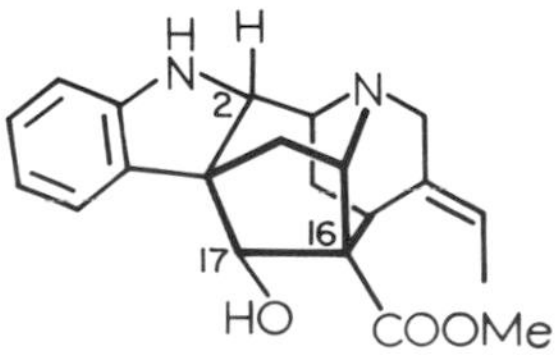

Quebrachidine 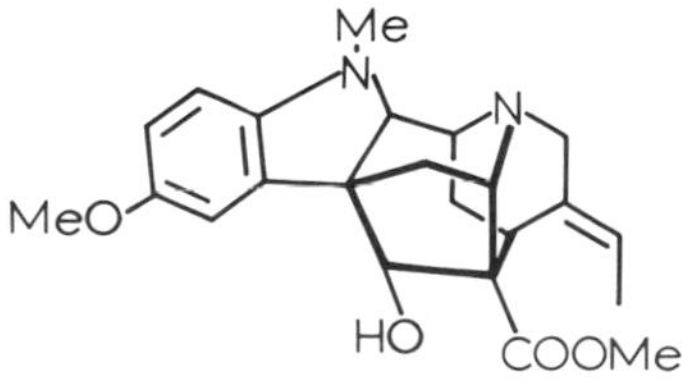   Vincamajoreine

It has been proposed that when the C-2 proton is cis to the
C-17 bridge it migrates to C-17 (path A, Chart 21) under electron impact induced cleavage generating a sarpagine-like

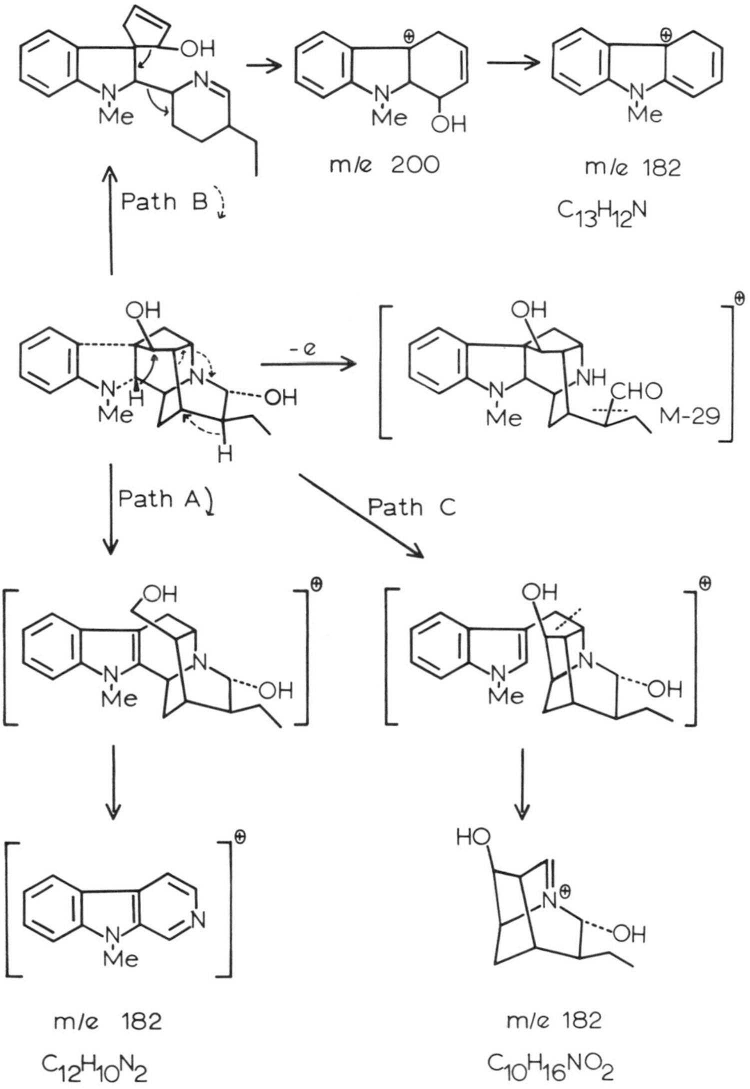

m/e 200

m/e 182

$C_{13}H_{12}N$

Path B

-e

M-29

Path A

Path C

m/e 182

$C_{12}H_{10}N_2$

m/e 182

$C_{10}H_{16}NO_2$

CHART 21 Some fragmentation products of ajmaline.

skeleton which then gives rise to peaks the equivalent to those derived from sarpagine itself. As would be expected, these peaks were absent from the spectra of 2-epi compounds where the C-2-C-3- bond is the one which moves (path C, Chart 21). This type of fragmentation was first observed with quebrachidine, its O, N-diacetate and vincamedine (1-methyl-quebrachidine-17-O-acetate).

E. Hervine, Herbaceine and Herbaine

From the mother liquors of the isolation of herbaceine and herbaine from V. herbacea, 7-methoxyindole alkaloid, hervine,

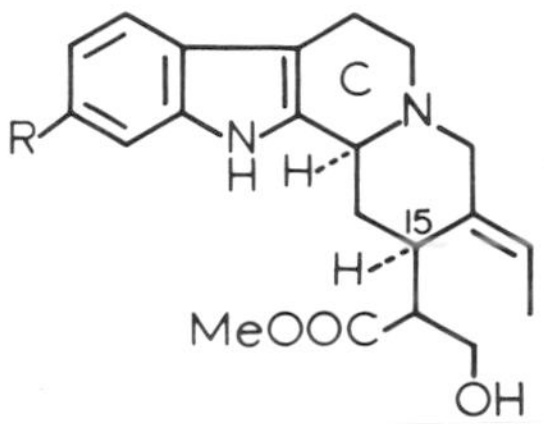

Hervine, R = OMe
Isositsirikine, R = H

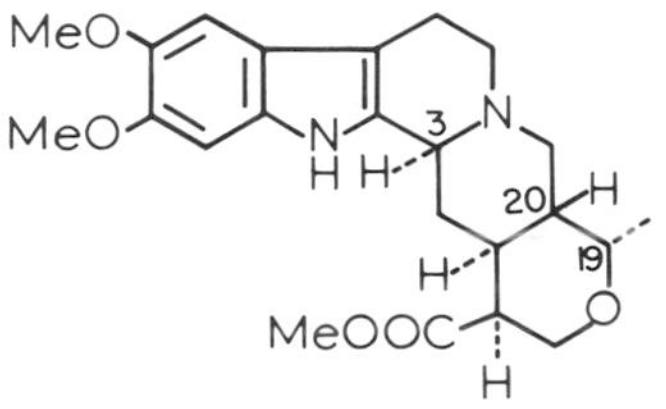

Herbaceine

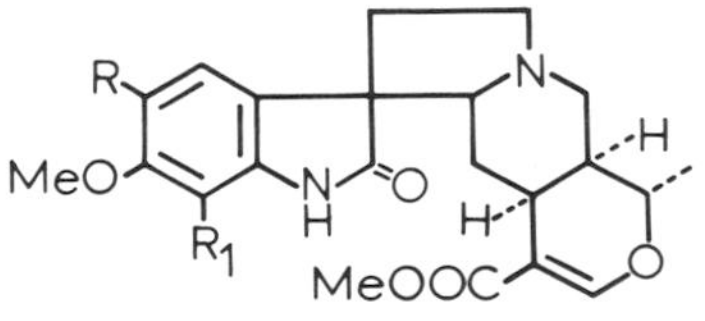

Reserpinine oxindole, R=H, R₁=H
Isoreserpiline oxindole, R=OMe, R₁=H
Majdine, R=H, R₁=OMe

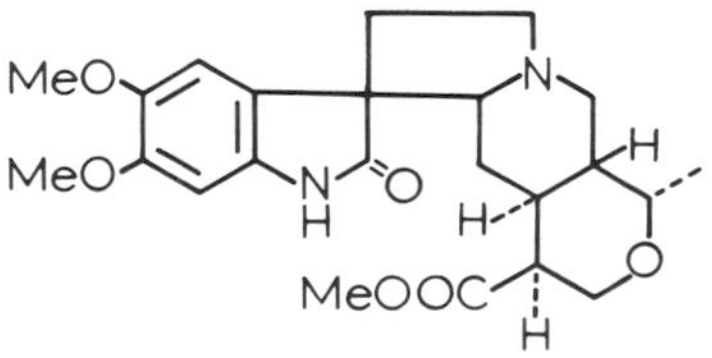

Herbaline

was isolated.[55] From an analysis of the physical data on her-

vine, O-acetylhervine and hervindiol(LiAlH$_4$ reduction product)

UV, IR, NMR and mass spectrum, it was thought to be a methoxy

isositsirikine. This was confirmed by a mass spectrometric

comparison of hervine with sitsirikine. The two spectra were

essentially identical if allowance was made for the 30 mass

unit difference which would appear in aromatic fragmentation

products of hervine. The C-3 stereochemistry follows from

its positive Cotton effect between 250-280 nm typical for 3α

yohimbans and corynantheans[33] and also mercuric acetate

oxidation of hervine to a ring C-tetradehydro derivative which

yields the starting product upon sodium borohydride reduction.

[55] The stereochemistry at C-15 is assumed to be α oriented

as are all known Type I alkaloids.

Herbaceine is an indole (mass spec. M=414.5) with three

methoxy groups, and is inert to hydrogenation in the presence

of Adams' catalyst.[56] Herbaceine has one active hydrogen, an

ind-NH as judged by the infrared spectrum and a >CH-CH$_3$

according to its NMR spectrum. From the mass spectrum and

the ultraviolet absorption spectrum herbaceine must be a 10-11-

dimethoxyindole [cf. seredine].[57] The mass spectral data in-

dicates that herbaceine is a complex dimethoxytetrahydro-β-

carboline[58] which is also consistent with its oxidation to 3-

dehydro-herbaceine (λ_{max} 310, 334 and 400 nm) and reversion to the starting alkaloid with sodium borohydride. The last experiment suggests that the CD ring is <u>trans</u>-fused (i. e. C-3 proton axial and <u>trans</u> to the lone pair orbital on N-4), a conclusion supported by the presence of the Bohlmann bands in the infrared (2700-2900 cm^{-1}) of herbaceine.

Hydrolysis of herbaceine yields methanol and epi-herbaceinic acid [hydrochloride, mp >200° (decomp.)] which upon esterification with diazomethane forms epiherbaceine. The epi-compound can also be produced by refluxing herbaceine in methanolic sodium methoxide for fifteen hours. Herbaceine upon reduction with lithium aluminum hydride furnishes herbaceinol. These experiments along with the inertness of herbaceine towards acylated agents led to the depicted structure stereochemistry at C-16, C-19 and C-20 still to be determined.

The base-catalyzed isomerization of herbaceine is best interpreted in terms of an axial to equatorial transformation for the methoxycarbonyl group. Pseudo first order rates of methiodide formation for herbaceine and 16-epiherbaceine show a marked similarity to those for corynanthine and yohimbine rather than to ring CD cis fused alkaloids. Spin decoupling of the C-19 hydrogen by irradiation of the C-19 methyl group indicated a dihedral angle of 60° hence a α-C-19

methyl. The chemical shift of the C-19 methyl compares well

with normal ring E-oxygen heterocyclic bases such as

ajmalicine. All of this evidence allows the full stereochemis-

try to be as indicated.[59]

Based on a limited amount of evidence, mainly of a physi-

cal nature, herbaine (vincaherbine[60]) is considered to be a 10-

demethoxyherbaceine.[56]

F. Oxindoles

Vinerine and vineridine from V. erecta seem to be

oxindoles isomeric with reserpinine oxindole.[61] Both bases

upon acetylation gave the same acetyl derivative, hydrolysis

of which regenerated vinerine. Reserpinine oxindole, was

prepared as follows: reserpinine with lead tetraacetate gave 7-

acetoxy-7H-reserpinine, which with acid rearranged into the

oxindole. Although the ultraviolet absorption spectrum of this

oxindole was in good agreement with those of vinerine and

vineridine, the respective infrared spectra were dissimilar.

It is believed that the difference lies in the configuration at C-

19 and/or C-20.

Majdine from V. major after treatment with acetic anhy-

dride and hydrolysis gives isomajdine, mp 204-206°, $[a]_D$ -90°

(MeOH). This transformation parallels that for vineridine to

vinerine and is typical for yohimbinoid oxindoles.[62] The

identification of carapanaubine (isoreserpiline oxindole A) in

V. pubescens[63] along with the close relationship between these

two species led to the suggestion that isomajdine might turn

out to be carapanaubine.[63] This conclusion was very nearly

correct since majdine has now been established to be the 11,

12-dimethoxy congener of carapanaubine.[74]

The evidence for this conclusion depends on a detailed

NMR comparison of majdine and isomajdine with known models.

It was not easy to exclude the 9, 10-dimethoxy possibility. Iso-

majdine is isomeric with majdine at C-7, the normal center

for isomerism among these oxindole alkaloids.

Herbaline from its ultraviolet absorption spectrum and

other data seems to be the oxindole analog of herbaceine[59, 64];

however, no attempt appears to have been made to make a

structure proof trivial by converting herbaceine into its oxin-

dole equivalent [cf. the conversion of reserpiline into cara-

panaubine].[65] A detailed examination of the resemblance be-

tween the NMR data of herbaceine and herbaline and the

principal peaks in the mass spectrum of herbaline are exactly

those which would have been predicted. In Table 4, the princi-

pal peaks obtained for herbaline and three other yohimbinoid

TABLE 4

Principal Fragments (m/e) in the Mass Spectra of Some Oxindole Alkaloids

Mitraphylline	130	144	145	146	159	208	223	353	368
Yohimbine oxindole B	130	144	145	146	159	207	225	-	370
Carapanaubine	190	204	205	206	219	208	223	413	428
Herbaline	190	204	205	206	219	210	225	415	430

oxindoles[66] are compared.[64] As expected, the parallelism
with carapanaubine is excellent.

G. Akuammine Group

The ease with which the structure of vincoridine (Chart 22)
was determined[67] was initially due to a good interpretation of
the physical data, especially the mass spectra which contained
peaks m/e 144 and 171 suggestive of an echitamine or corymine
type of skeleton.[68]

Reduction of vincoridine with sodium borohydride yielded
a compound whose chromatographic mobility and mass spectrum
was identical with diformylcorymine XXXV (treatment of cory-

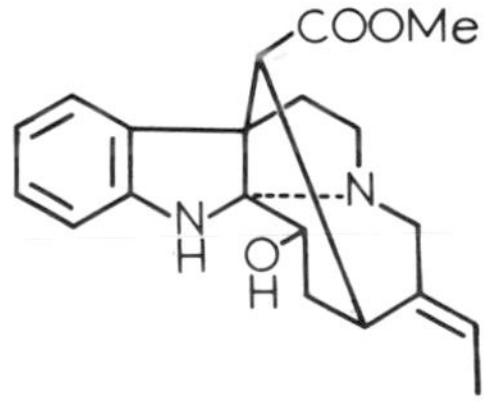

XXXV

Deformylcorymine

mine with hot aqueous potassium hydroxide).[69] The absolute
stereochemistry depicted (Chart 22) has not been unequivocally
established but is probably correct.

There is probably a close biogenetic relationship between
vincoridine and the strictamine group shown in Chapter 2, Fig.

11. Vincamidine and vincamajoridine were recognized to be identical with strictamine and akuammine, respectively. The structures for ervincine[75] and vincaridine[76] are suggested ones and remain to be proven.

H. Akuammicine Group

The chemistry of this group is closely identified with that

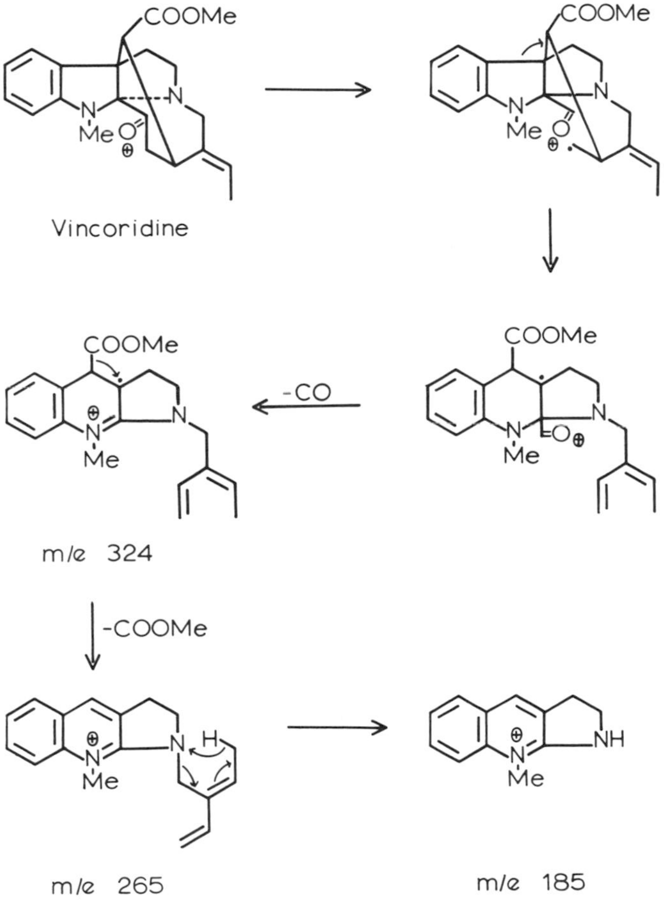

CHART 22 Fragmentation of vincoridine under electron impact.

developed for norfluorocurarine and akuammicine. The known members are listed in Chapter 2, Fig. 8, and their interrelationships are clear.

The chemistry of the phenolic base vinervine (XXXVI) parallels that of akuammicine. With sodium borohydride it yields a 2, 16-dihydro derivative, ozonolysis gives acetaldehyde, and in 15% hydrochloric acid at 100° it furnishes demethoxy-carbonylvinervine.[70] Vinervine forms an O-methyl ether, with diazomethane and this product after treatment with 20% hydrochloric acid in a sealed tube at 110° followed by sodium borohydride reduction (in aqueous methanolic hydrochloric acid) gave the same indoline, mp 240-242°, as that formed from vincanidine by an analogous procedure.[71] The phenolic hydroxyl was placed at C-11 since the ultraviolet absorption spectrum

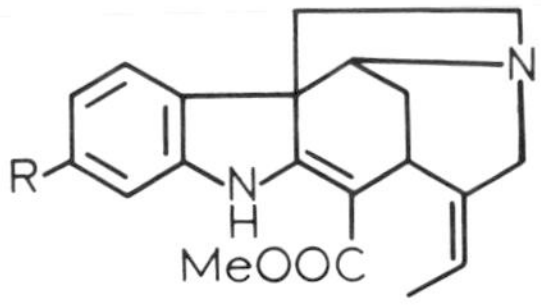

XXXVI

Akuammicine, R = H

Vinervine,   R = OH

of 2, 16-dihydrovinervine was found to closely resemble model 11-hydroxytetrahydro-β-carboline derivatives[71] and this was proven by the oxidation of deformyl-O-methyldihydrovincanidine to 4-methoxy N-oxalylanthranilic acid.[72] The overall structure was finally established by conversion of O-tosyltetrahydro-vinervine into tetrahydroakuammicine by Raney nickel.[71,72]

REFERENCES

1. J. Le Men and W. I. Taylor, Experientia, 21, 508 (1965).

2. E. Schlittler and W. I. Taylor, Experientia, 16, 244 (1960).

3. J. Trojánek, O. Štrouf, J. Holubek and Z. Čekan, Tetrahedron Letters, 1961, 702; Collection Czech. Chem. Commun., 29, 433 (1964).

4. J. Mokrý, I. Kompiš and P. Šefčovič, Tetrahedron Letters, No. 10, 433 (1962).

5. O. Štrouf and J. Trojánek, Chem. Ind. (London), 1962, 2037; Collection Czech. Chem. Commun., 29, 447 (1964).

6. M. Plat, D. D. Manh, J. Le Men, M. -M. Janot, H. Budzikiewicz, J. M. Wilson, L. J. Durham and C. Djerassi, Bull. Soc. Chim. France, 1962, 1082.

7. R. T. Major and I. El Kohly, J. Org. Chem., 28, 591 (1963).

8. J. Mokrý, I. Kompiš, and P. Šefčovič, Tetrahedron Letters, 1962, 433.

9. M. Plat, R. Lemay, J. Le Men, M. -M. Janot, C. Djerassi and H. Budzikiewicz, Bull. Soc. Chim. France, 1965,

10. J. Mokrý, M. Shamma and H. E. Soyster, Tetrahedron Letters, 1963, 999.

11. E. Wenkert, B. Wickberg and C. Leicht, Tetrahedron Letters, 1961, 822.

12. J. Mokrý and I. Kompiš, Tetrahedron Letters, 1963, 1917.

13. J. Mokrý and I. Kompiš, Lloydia, 27, 428 (1964).

14. J. Trojánek, Z. Kablicová and K. Bláha, Chem. Ind. (London), 1965, 1261.

15. M. E. Kuehne, J. Am. Chem. Soc., 86, 2946 (1964); Lloydia, 27, 435 (1964).

16. J. Holubek, O. Štrouf, J. Trojánek, A. J. Bose and E. R. Malinowski, Tetrahedron Letters, 1963, 897.

17. B. Pyuskyulev, I. Kompiš, I. Ognyanov and G. Spiteller, Collection Czech. Chem. Commun., 32, 1289 (1966).

18. M. F. Bartlett, W. I. Taylor and Raymond-Hamet, Compt. Rend. Acad. Sci., 249, 1259 (1959).

19. M. F. Bartlett and W. I. Taylor, J. Am. Chem. Soc., 82, 5941 (1960).

20. M. F. Bartlett and W. I. Taylor, Tetrahedron Letters, No. 20, 20 (1959).

21. J. Mokrý, I. Kompiš and P. Šefčovič, Tetrahedron Letters, No. 10, 433 (1962).

22. E. Wenkert and B. Wickberg, J. Am. Chem. Soc., 87, 1580 (1965).

23. J. E. D. Barton and J. Harley-Mason, Chem. Commun., 1965, 298; J. E. D. Barton, J. Harley-Mason and K. C. Kates, Tetrahedron Letters, 1965, 3669.

24. G. Stork and J. E. Dolfini, J. Am. Chem. Soc., 85, 2872 (1963).

25. H. K. Schnoes, A. L. Burlingame and K. Biemann, Tetrahedron Letters, 1962, 993.

26. L. D. Antonaccio, N. A. Pereira, B. Gilbert, H. Vorbrueg-gen, H. Budzikiewicz, J. M. Wilson, L. J. Durham and C. Djerassi, J. Am. Chem. Soc., 84, 2161 (1962).

27. J. F. D. Mills and S. C. Nyburg, J. Chem. Soc., 1960, 1458.

28. G. F. Smith and J. T. Wróbel, J. Chem. Soc., 1960, 1463.

29. H. Kny and B. Witkop, J. Org. Chem., 25, 635 (1960).

30. B. Witkop, J. Am. Chem. Soc., 70, 3712 (1948); 79, 3193 (1957).

31. F. Walls, O. Collera and A. Sandoval, Tetrahedron, 2, 173 (1958).

32. K. Biemann and G. Spiteller, J. Am. Chem. Soc., 84, 4578 (1962).

33. D. Schumann, B. W. Bycroft and H. Schmid, Experientia, 20, 202 (1964); B. W. Bycroft, D. Schumann, M. B. Patel and H. Schmid, Helv. Chim. Acta, 47, 1147 (1964); W. Klyne, R. J. Swan, B. W. Bycroft, D. Schumann and H. Schmid, Helv. Chim. Acta, 48, 443 (1965); W. Klyne, R. J. Swan, N. J. Dastoor, A. A. Gorman and H. Schmid, Helv. Chim. Acta, 50, 115 (1967).

34. J. Mokrý, I. Kompiš, L. Dúbrakova and P. Šefčovič, Tetrahedron Letters, 1962, 1185.

35. J. Trojánek, O. Štrouf, K. Bláha, L. Dolejš and V. Hanuš, Collection Czech. Chem. Commun., 29, 1904 (1963).

36. J. P. Kutney, K. K. Chan, A. Failli, J. M. Fromson, C. Gletsos and V. R. Nelson, J. Am. Chem. Soc., 90, 3893 (1968).

37. J. Mokrý, I. Kompiš, M. Shamma and R. J. Shine, Chem. Ind. (London), 1964, 1988.

38. M. Plat, J. Le Men, M.-M. Janot, J. M. Wilson, H. Budzikiewicz, L. J. Durham, Y. Nakagawa and C. Djerassi, Tetrahedron Letters, 1962, 271.

39. C. Djerassi, H. Budzikiewicz, J. M. Wilson, J. Gosset, J. Le Men and M.-M. Janot, 1962, 235.

40. K. Biemann, M. Friedmann-Spiteller and G. Spiteller, Tetrahedron Letters, 1961, 485.

41. M. Plat, J. Le Men, M. -M. Janot, H. Budzikiewicz, J. M.
    Wilson, L. J. Durham and C. Djerassi, Bull. Soc. Chim.
    France, 1962, 2237.

42. V. M. Malikov, P. Kh. Yuldashev and S. Yu. Yunusov, Dokl.
    Akad. Nauk. Uz. SSR., 20, No. 4, 21 (1963).

43. W. I. Taylor in "The Alkaloids" (R. H. F. Manske, ed.)
    p. 785, Academic Press, New York, 1965.

44. M. F. Bartlett, R. Sklar, W. I. Taylor, E. Schlittler, R. L.
    S. Amai, P. Beak, N. V. Bringi and E. Wenkert, J. Am.
    Chem. Soc., 84, 622 (1962).

45. E. Wenkert and N. V. Bringi, J. Am. Chem. Soc., 81,
    1474, 6535 (1959); E. Ochiai and M. Ishikawa, Pharm. Bull.
    (Japan) 6, 208 (1958); V. Prelog and E. Zalán, Helv. Chim.
    Acta, 27, 545 (1944).

46. S. Masamune, S. K. Ang, C. Egli, N. Nakatsuka, S. K.
    Sarkar and Y. Yasunari, J. Am. Chem. Soc., 89, 2506
    (1967).

47. S. Silvers and A. Tulinsky, Tetrahedron Letters, 1962,
    339; A. T. McPhail, J. M. Robertson, G. A. Sim, A. R.
    Battersby, H. F. Hodson and D. A. Yeowell, Proc. Chem.
    Soc., 1961, 223.

48. M. F. Bartlett, B. F. Lambert, H. M. Werblood and W. I.
    Taylor, J. Am. Chem. Soc., 85, 475 (1963).

49. M. Gorman, A. L. Burlingame, and K. Biemann, Tetra-
    hedron Letters, 1963, 39.

50. P. Kh. Yuldashev and S. Yu. Yunusov, Khim. Prirodn.
    soedinen Akad. Nauk SSR, No. 2, 110 (1965); Dokl. Akad.
    Nauk SSR, 163, No. 1, 123 (1965).

51. K. Biemann, P. Bommer, A. L. Burlingame and W. J.
    McMurray, J. Am. Chem. Soc., 86, 4624 (1964);
    Tetrahedron Letters, 1963, 1969.

52. M. Plat, R. Lemay, J. LeMen, M. -M. Janot, C. Djerassi
    and H. Budzikiewicz, Bull. Soc. Chim. France, 1965,
    2497; P. Potier, R. Beugelmans, J. Le Men and M. -M.
    Janot, Ann. Pharm. Franc., 23, 61 (1965).

53. J. L. Kaul, J. Trojánek and A. K. Bose, Chem. Ind. (London), 1966, 853.

54. M. Falco, J. Garnier-Gosset, E. Fellion and J. Le Men, Ann. Pharm. France, 22, 455 (1964).

55. I. Ognyanov, B. Pyuskyulev, B. Bozjanov and M. Hesse, Helv. Chim. Acta, 50, 754 (1967).

56. I. Ognyanov, and B. Pyuskyulev, Ber., 94, 1008 (1966).

57. J. Poisson, N. Neuss, R. Goutarel and M. -M. Janot, Bull. Soc. Chim. France, 1958, 1195.

58. H. Budzikiewicz, C. Djerassi and D. H. Williams, "Structure Elucidation of Natural Products, Vol. I, Alkaloids", Holden Day, San Francisco, 1964.

59. I. Ognyanov, B. Pyuskyulev, M. Shamma, J. A. Weiss and R. J. Shine, Chem. Commun., 1967, 579.

60. E. S. Zabolotnaya and E. V. Bukreeva, Zh. Obshch. Khim., 33, 3780 (1963).

61. Sh. Z. Kasymov, P. Kh. Yuldashev and S. Yu. Yunusov, Dokl. Akad. Nauk. SSSR, 162 (1), 102 (1965); 163 (6), 1400 (1965).

62. J. C. Seaton, M. D. Nair, O. E. Edwards and L. Marion, Canad. J. Chem., 38, 1035 (1960).

63. N. Abdurakhimova, P. Kh. Yuldashev and S. Yu. Yunusov, Khim. Prirodn. Soedn. Akad. Nauk. Uz. SSR, No. 3, 224 (1965).

64. I. Ognyanov, Ber., 99, 2052 (1966).

65. N. Finch, C. W. Gemenden, I. H. Hsu and W. I. Taylor, J. Am. Chem. Soc., 85, 1520 (1963).

66. B. Gilbert, J. A. Brissolese, N. Finch, W. I. Taylor, H. Budzikiewicz, J. M. Wilson, and C. Djerassi, J. Am. Chem. Soc., 85, 1523 (1963).

CHAPTER 4

CHEMOTAXONOMY OF VINCA SPECIES

J. Trojánek, M. Nováček and F. Starý

Research Institute for Pharmacy and Biochemistry,
Prague, Czechoslovakia

Previous chapters of this monograph have clearly demonstrated the enormous enrichment of our knowledge of the genus
Vinca during the past 10-15 years. It might seem, therefore,
somewhat surprising that the taxonomic values of certain taxa
and/or their position within this genus are not yet fully recognized, and as such are sources of many disputes. This can be
seen from Table 1 where taxonomic considerations proposed
by some recent botanists are given. For instance, Pichon[1]
considers V. erecta as synonymous with V. herbacea, but other
authorities[2,3] consider these two species as completely independent. Moreover, Pobedimova[2] places V. erecta as the only
representative of a new section, i. e. Vincopsis Pobedimova.
Also, the taxonomic value of V. pubescens with relationship to
V. major, is confusing. Similar confusion exists relative to

TABLE 1

Botanical Classification of the Genus Vinca According to

Pichon (1)	Pobedimova (2)	Stearn (3)
V. herbacea Waldst. et Kit. (syn. V. erecta Regel et Schmalh.)	V. herbacea Waldst. et Kit.	V. herbacea Waldst. et Kit.
var. herbacea		
var. libanotica (Zucc.) M. Pichon		
var. sessilifolia (A. DC.)		
	V. erecta Regel et Schmalh.	V. erecta Regel et Schmalh.
V. major L.	V. major L.	V. major L.
var. major (syn. V. pubescens Urv.)	V. pubescens Urv.	subsp. major
		subsp. hirsuta (Boiss.) Stearn (syn. V. pubescens Urv.)
var. difformis (Pourr.) M. Pichon		V. difformis Pourr.
		V. sardoa Stearn
		V. libanotica Penzes
V. minor L.	V. minor L.	V. minor L.
var. minor		
var. nammulariaefolia Fournier		

V. difformis, V. major[1,3] and other species. There is little
doubt that some of these uncertainties arise due to a limited
accessibility to suitable living plant material originating from
Middle Asia and Asia Minor (Turkey, Lebanon and adjacent regions). Asia Minor is considered to be the possible center of
the development of the genus Vinca. It is probably dangerous
to consider in too critical a manner, intraspecific taxonomic
evaluations of individual taxa in this group until detailed chorologic and karyologic data become available.

During the past decade many experiments were undertaken
with the aim to clarify some of these questions from a chemical
point of view; or, to be more exact, from the point of view of
the occurrence of specific constituents, which might be characteristic for certain taxa. Thus, Paris and Moyse-Mignon[4]
were the first who approached the problem in this manner.
They found that the electrophoretograms of alkaloid complexes
obtained from V. major and V. difformis were basically different,
and in this way implied serious doubt on the correctness of
Pichon's taxonomic concept[1], according to which both species
should represent varieties of the same species V. major. Analogous differences were observed[4] also in the case of alkaloid
complexes obtained from similarly related[1] taxa, i.e. V. herbacea and V. libanotica (Table 1).

In 1965, Janot, Le Men and Garnier[5] published a more de-

tailed chemotaxonomic study based on the distribution of three

basic classes[6] of indole alkaloids - A, B and C - characterized

by the structure of the aliphatic C_{10}-unit (now known to be of

terpenoid origin)[7], and by the distribution of dimeric alkaloids

of a mixed B-C class in the taxa of the genus Vinca, and of the

closely related subtropical genus Catharanthus. They found,

in agreement with earlier observations[4], basic chemical dif-

ferences between these two genera, thus corroborating Pichon's[8]

taxonomic opinion. Further, they pointed out that the occur-

A   B   C

(Yohimbanoids)   (Aspidospermanoids)   (Iboganoids)

rence of alkaloids of class C(iboganoids) and of dimeric bases

of the mixed B-C classes, seems to be characteristic for the

genus Catharanthus, because they were not identified in any

Vinca taxon studied up to that time. They also turned attention

to the fact that among alkaloids of V. minor only bases of class

B (aspidospermanoids) were found, while in other taxa of Vinca,

aspidospermanoids (class B) together with yohimbanoids (class

A), always co-occur, the latter prevailing regularly. Finally,
on the basis of comparative studies of the alkaloids of V. major
and V. difformis, they concluded that the differences in the composition of the alkaloidal constituents of these species were so
pronounced that they were incompatible with Pichon's theory of
the intraspecific relationship of both taxa[1]. Probably, in this
case, chemical criteria should be considered as being more
important than botanical criteria[5].

Since that paper was published, the number of alkaloids of
known structure isolated from different taxa of Vinca has increased considerably. It seems, therefore, reasonable to dis-
cuss newly accumulated chemical data with their relationship
to existing taxonomic systems.

Recently, based on the results of their broad studies of rich
herbarium material, and on a critical evaluation of published
chorologic data, Nováček and Starý[9] divided the genus Vinca
into four sections, including a total of eight species (Table 2).
As will be shown later, this classification agrees well, not only
with the distribution of the basic alkaloid classes A and B
(Table 4), but also with that of corresponding structural groups
within these classes (Table 5). For better orientation, all
alkaloids of known structure found in the individual taxa of the
genus Vinca are presented in Table 3. The alkaloids are divid-

TABLE 2

Classification of Vinca Species According

to Nováček and Starý (9)

Section	Species
1. Minor	1. V. minor L.
2. Major	2a. V. major L.
	2b. V. difformis Pourr.
	2c. V. pubescens Urv.
3. Herbacea	3a. V. herbacea Waldst. et Kit.
	3b. V. libanotica Zucc.
	3c. V. haussknechtii Bornm. et Sint.
4. Erecta	4. V. erecta Regel et Schmalh.

ed arbitrarily into classes A and B, and further within these classes into several subgroups according to their increasing complexity and structural relationship.

Initially, we wish to point out that some of these structures are not rigorously established and could be incorrect. Recently, for example, it has been shown[10] that the alkaloid ervincine, originally thought[11] to possess a picrinine-type structure, does

TABLE 3

Distribution of Indole Alkaloids in <u>Vinca</u> Species

CLASS A

Group A-1.1

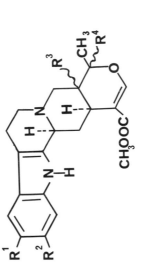

	R[1]	R[2]	R[3]	R[4]	Occurrence[a]
Ervine (= Rauniticine) (?)	H	H	αH	αH	4
Ajmalicine	H	H	βH	βH	4
Reserpinine	H	OCH₃	αH	βH	4
Isoreserpiline (10)	OCH₃	OCH₃	αH	βH	2a, 2c, 3a, 4

a   According to Table 2.

Group A-1.2

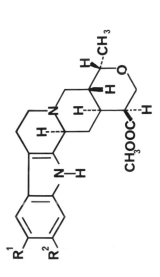

	R^1	R^2	Occurrence^a
Herbaceine	OCH$_3$	OCH$_3$	3a
Herbaine^b	H	OCH$_3$	3a

^{a} According to Table 2.
^{b} Stereochemistry at C$_{(15)}$, C$_{(16)}$, C$_{(19)}$ and C$_{(20)}$ unknown

Group A-2.1

CLASS A

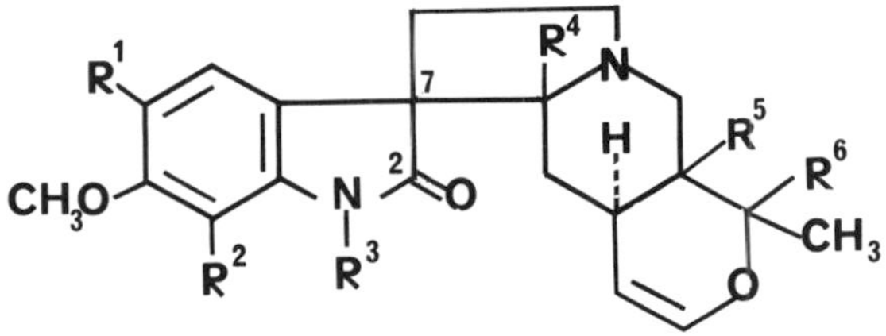

	R^1	R^2	R^3	R^4	R^5	R^6	C7(2)	Occurrence
Ercinine(12)	H	H	H	αH	αH	βH	?	4
Vinerine	H	H	H	?H	?H	?H	?	4
Vineridine	H	H	H	?H	?H	?H	?	4
Carapanaubine	OCH$_3$	H	H	αH	αH	βH	β	2c
Majdine	H	OCH$_3$	H	αH	αH	βH	β	2a, 3a
Isomajdine	H	OCH$_3$	H	αH	αH	βH	α	3a
N-Acetylvinerine	H	H	COCH$_3$	αH	?H	?H	?	4

CLASS A

Group A-3

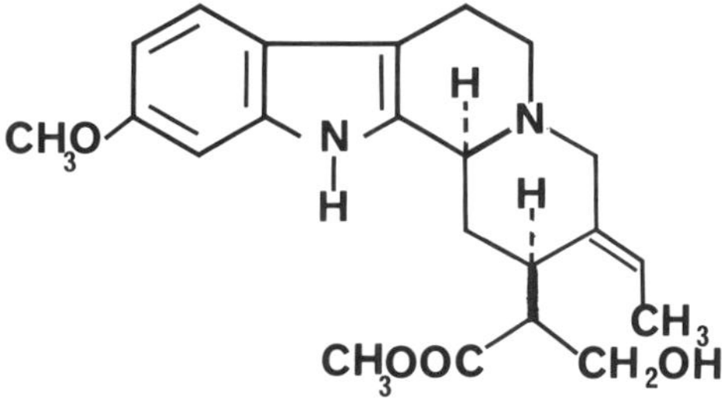

Hervine 3a

Group A-2.2

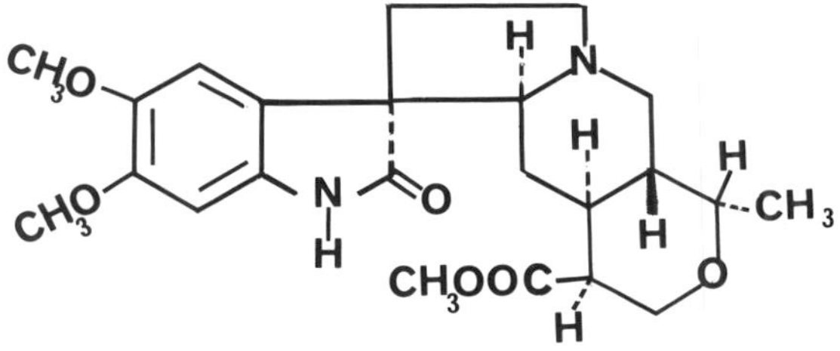

Herbaline 3a

CLASS A

Reserpine

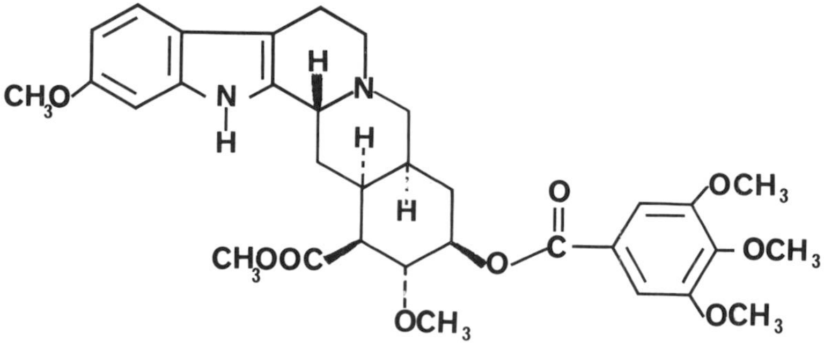

Group A-4.2

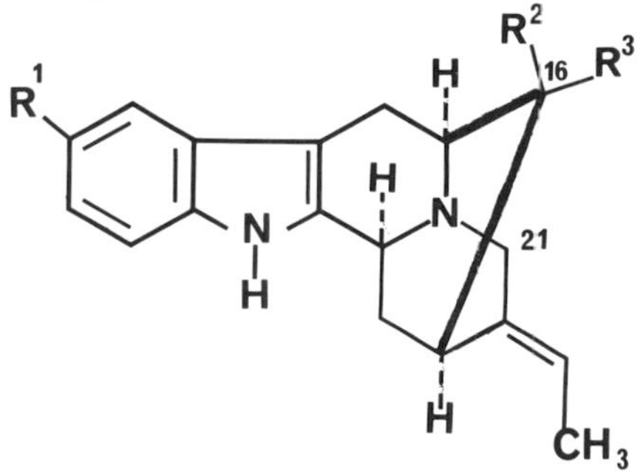

Group A-4.1

	R¹	R²	R³	
		COOCH$_3$	CH$_2$OH	
Akuammidine	H	COOCH$_3$	CH$_2$OH	2b, 4
Vellosimine	H	H	CHO	2b
Tombozine	H	H	CH$_2$OH	4
Sarpagine	OH	H	CH$_2$OH	2b
10-Methoxyvelosimine	OCH$_3$	H	CHO	2a
Ervincidine	16- or 21-hydroxytombozine (10) (?)			

CLASS A

Group A-4.3

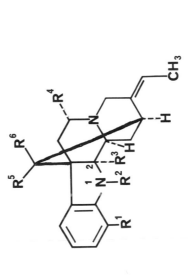

	R^1	R^2	R^3	R^4	R^5	R^6	Occurrence
Picrinine (= vincaridine)	H	H	--0--		$COOCH_3$	H	4
Vincaricine	OCH_3	H	--0--		$COOCH_3$ (Structure?)	H	4
Vincamidine (= stricatamine)	H	-	-	H	H	$COOCH_3$	1 (1,2-double bond)

CLASS A

Group A-4.4

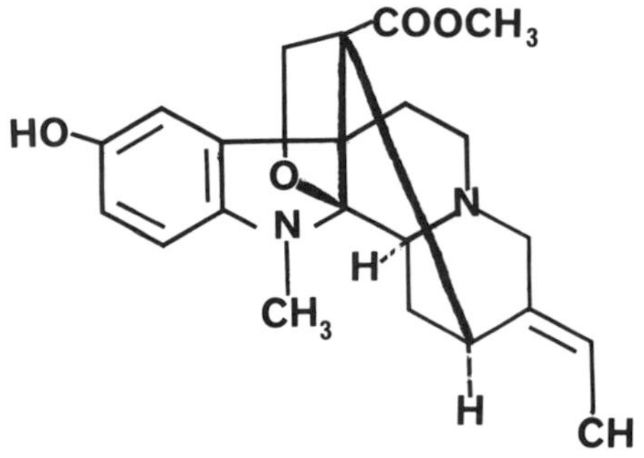

Akuammine (= vincamajoridine)   2a, 4

Group A-4.5

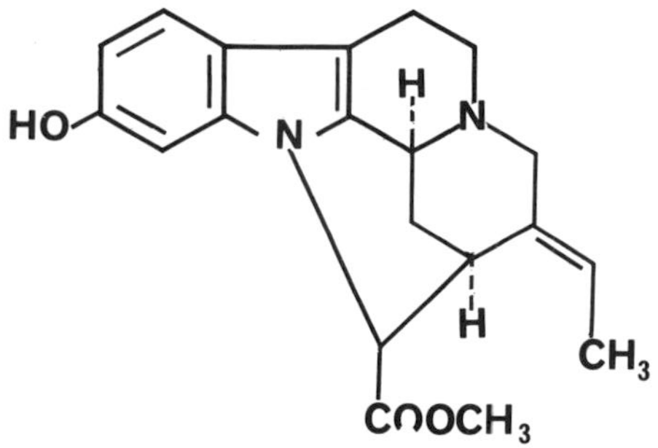

11-Hydroxypleiocarpamine (13)   4

CLASS A

Group A-5

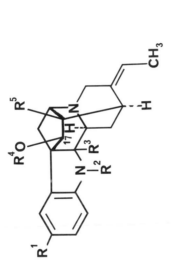

	R^1	R^2	R^3	R^4	R^5	Occurrence
Vincarine	H	H	?H	H	COOCH$_3$ (configuration at C-17 ?)	4
Vincamedine	H	CH$_3$	αH	COCH$_3$	COOCH$_3$	2a, 2b
Vincamajine	H	CH$_3$	αH	H	COOCH$_3$	2a, 2b
Majoridine	H	CH$_3$	βH	COCH$_3$	H	2a
Vincamajoreine	OCH$_3$	CH$_3$	βH	H	H	2a

CLASS A

Group A-6

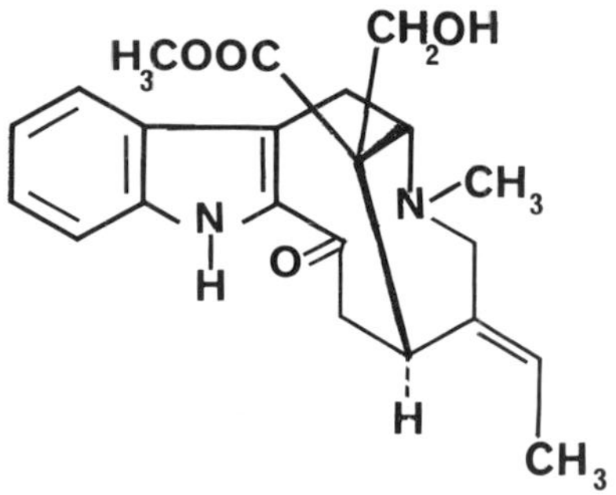

Vincadaffine

2b

Group A-7

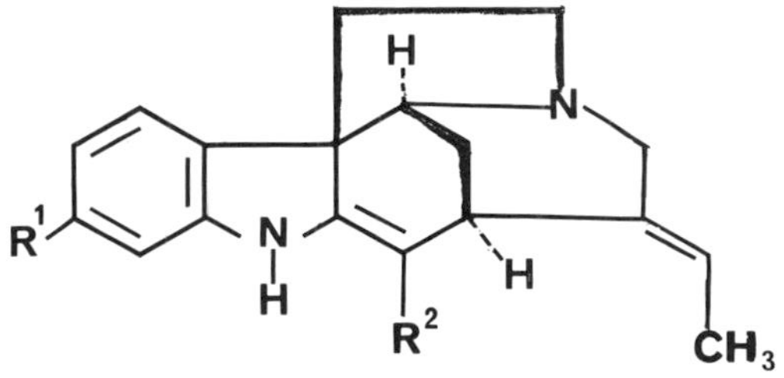

	R^1	R^2	Occurrence
Vincanine (= norfluorocurarine)	H	CHO	3a, 4
Vincervidine (= (+)-vincanine) (14)	H	CHO	4
Vincanidine	OH	CHO	4
Vincanicine (14)	OCH_3	CHO	4
Akuammicine	H	$COOCH_3$	4
Vinervine	OH	$COOCH_3$	4
Vinervinine	OCH_3	$COOCH_3$	4

Group A-8

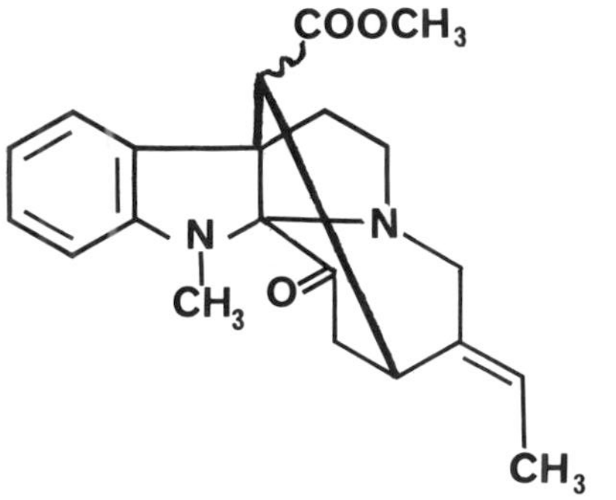

Vincoridine   1

CLASS B

Group B-1

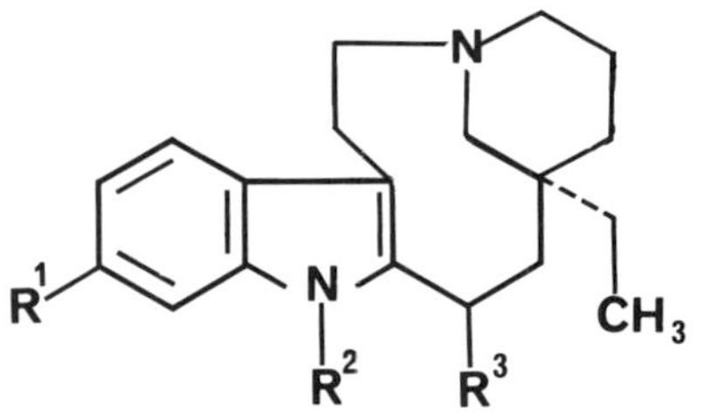

	R¹	R²	R³	Occurrence 1, 4 (10)
(+)-Quebrachamine	H	H	H	1
(±)-ind-N-methylque-brachamine	H	CH$_3$	H	1
Vincadine	H	H	α COOCH$_3$	1
Vincaminorine	H	CH$_3$	α COOCH$_3$	1
Vincaminoreine	H	CH$_3$	β COOCH$_3$	1
Vincaminoridine	OCH$_3$	CH$_3$	α COOCH$_3$	1

CLASS B

Group B-2.1

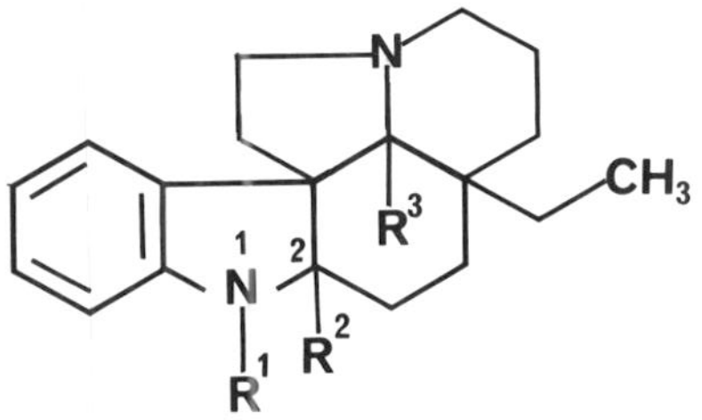

	R^1	R^2	R^3	C19(20)	C6(7)	
(+)-N-methylaspidospermidine	CH_3	αH	βH	β	α	1
(+)-1.2-Dehydroaspidospermidine	1.2-double bond		βH	β	α	1
(-)-1.2-Dehydroaspidosperimidine (10)	1.2-double bond		αH	α	β	1

CLASS B

Group B-2.2 (cont.)

	R¹	R²	R³	R⁴	R⁵	R⁶	C6(7)	C19(20)	Occurrence
16-Methoxytabersonine	OCH₃	H	H	H	H	H	β	α , 14.15-double bond	3a
Ervamicine[d]	OCH₃	H	=0	H	H	H	β?	α?, 14. 15-double bond	4
Minovincine	H	H	=0	H	H	H	β	α	1
(-)-Minovincinine	H	OH	H	H	H	H	β	α	1
Vincesine (16)	H	H	H	OH	H	H	β	α	1
16-Methoxyminovincine	OCH₃	H	=0	H	H	H	β	α	1
16-Methoxyminovincinine	OCH₃	OH	H	H	H	H	β	α	1
Ervinidine[e]	H	H	H	H	-0-		β?	α?	4
Lochnerinine	OCH₃	H	H	H	-0-		β	α	3a
Ervincinine[f]	OCH₃	H	H	H	-0-		?	α	4

---

d Probably an isomer of 16-methoxytabersonine (15)

e Structure ?? ervinidinine (17) should differ from lochnericine in the configuration at C(21)

f Structure ?? Ervincine (15) should differ from lochnerinine in the configuration at C(21)

CLASS B

Group B-2.2

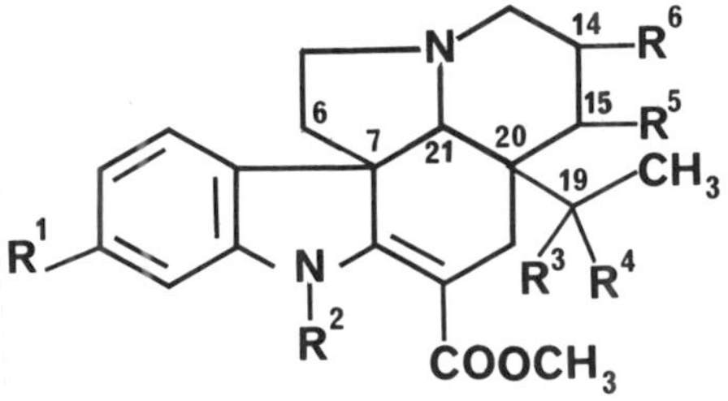

	R¹	R²	R³	R⁴	R⁵	R⁶	C6(7)	C19(20)	Occurrence
(-)-Vincadifformine	H	H	H	H	H	H	β	α	1
Ervamine [(-)-vinca-difformine?]	H	H	H	H	H	H	β	α	4
(+)-Vincadifformine	H	H	H	H	H	H	α	β	2b
(+)-Vincadifformine	H	H	H	H	H	H	-	-	1, 2b
(+)-Minovine	H	CH₃	H	H	H	H	-	-	1
16-Methoxyvinca-difformine	OCH₃	H	H	H	H	H	β	α	1
Ervinceine^c	OCH₃	H	H	H	H	H	β	α	4
Tabersonine	H	H	H	H	H	H	β	α, 14, 15-double bond	3a

c Probably an isomer of methoxyvincadifformine (15)

CLASS B

Group B-3.1

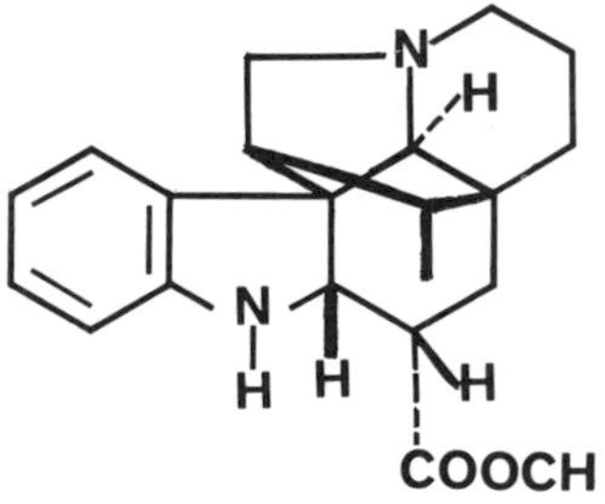

Pseudokopsinine

Occurrence
4

Group B-3.2

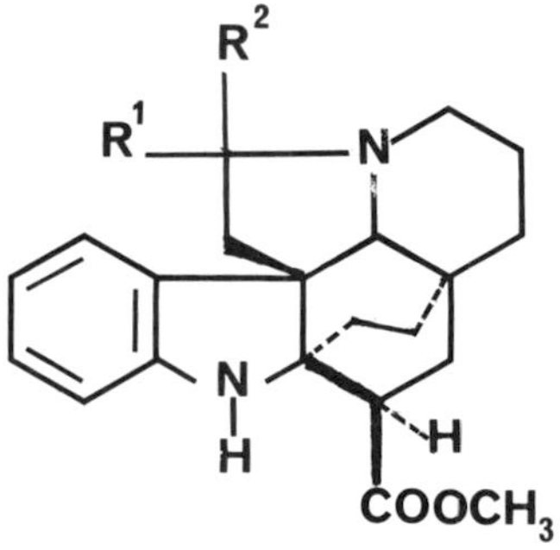

	R^1	R^2	
Kopsinine (= erectine)	H	H	4
Kopsinilam	=0		4

CLASS B

Group B-4

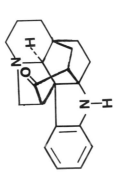

Kopsanone[g] (10)

[g] The physical constants given for kopsanone from Vinca erecta[10] are distinctly different from those of the authentic alkaloid[18]. On the other hand, they are very close to those of 10,22-dioxokopsane[18], so that both alkaloids may be identical.

CLASS B

Group B-5.1

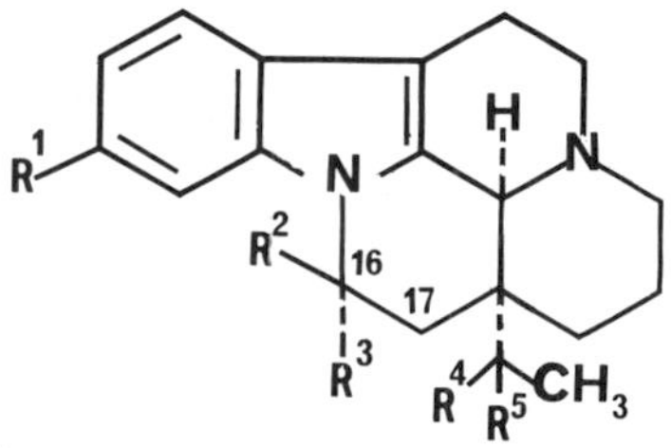

	R¹	R²	R³	R⁴	R⁵	Occurrence
Vincamine	H	OH	COOCH₃	H	H	1, 2a, 2b, 3a, 4
16-Epivincamine	H	COOCH₃	OH	H	H	1
Vincine	OCH₃	OH	COOCH₃	H	H	1, 2a, 4
Vincaminine	H	OH	COOCH₃	=O		1
Vincinine	OCH₃	OH	COOCH₃	=O		1
19-Hydroxyvincamine	H	OH	COOCH₃	OH	H (stereochemistry ?)	1
(–)-Eburnamonine	H	=O	=O	H	H	1, 4 (10)
(±)-Eburnamonine (= vincanorine)	H	=O	=O	H	H	1
(–)-11-Methoxyebur-namonine	OCH₃	=O	=O	H	H	1
Apovincamine	H	H	COOCH₃	H	H 16.17-double bond	4

h Artefact?

CLASS B

Group B-5.2

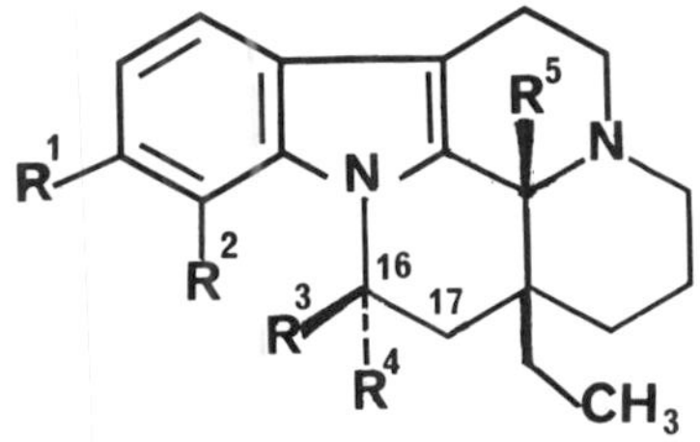

	R^1	R^2	R^3	R^4		Occurrence
Eburnamine	H	H	H	OH		1, 4 (10)
(+)-Eburnamine[1] (10)	H	H	H	OH		4
Isoeburnamine	H	H	OH	H		1
Eburnamenine	H	H	H	H	- 16, 17-double bond	1
11, 12-Dimethoxyebur- namonine	OCH$_3$	OCH$_3$	H	H	=0 (stereochemistry?)	1

[1] May well be a mixture of (+)-eburnamine and (-)-isoeburnamine

CLASS B

Group B-6

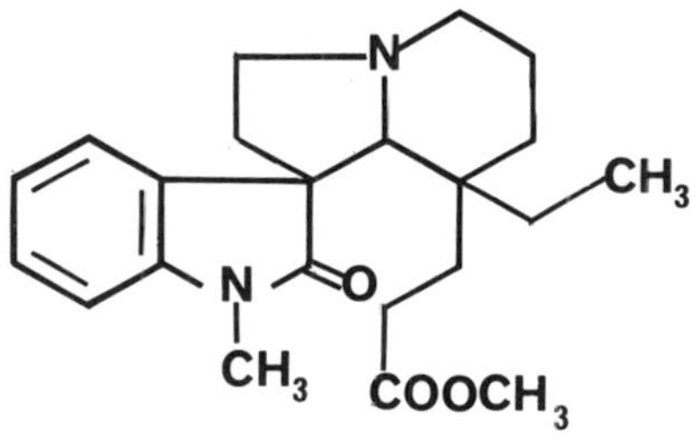

Vincatine

Occurrence

1

not belong in the indoline group of bases at all, but represents

a mixture of the known indole alkaloids vincamine and apovin-

camine. Moreover, it is not even clear whether apovincamine

is a genuine base, or an artefact arising from vincamine. Some

of the figures given in Tables 4 and 5 might, therefore, be

slightly misrepresented. In all cases where we were not ab-

solutely certain about the correctness of occurrence and/or

structure of a given alkaloid, we have placed a question mark

(?), or added a footnote at the appropriate place. Possible

misrepresentation of our conclusions may be further due to

the fact that we have neither considered the phytogeographical

characteristics of a plant which was chemically studied, nor

specified its anatomic ontogenic state, etc. But, we think that

for the purposes of comparative chemotaxonomy, at least for

an initial approach, it was sufficient to recognize the ability

of the plant under study to synthesize and accumulate alkaloids

of certain types.

Table 4 shows that the original chemotaxonomic findings

reported by Janot et al.[5] are still valid. The species V. minor

differs clearly from all other members of the genus Vinca in

its ability to synthesize more than 90 per cent of its alkaloids

of the class B type, while the remaining species which have

been studied up to the present time always produce alkaloids

TABLE 4

Distribution of Classes A and B of Indole Alkaloids in <u>Vinca</u> Species

Section	Species	Total Number of Alkaloids Described	Number of Structurally Known Alkaloids	Class A		Class B		Predominant Class
				Number	%	Number	%	
Minor	<u>V.minor</u>	40	33 (34a)	2 (3a)	6 (9a)	31	94 (91a)	B
Major	<u>V.major</u>	13	10	8	80	2	20	A
	<u>V.difformis</u>	10	9	6	67	3	33	A
	<u>V.pubescens</u>	3	2	2	100	0	0	A
Herbacea	<u>V.herbacea</u>	15	12	8	67	4	33	A
	<u>V.libanotica</u>	-	-	-	-	-	-	?
	<u>V.haussknechtii</u>	-	-	-	-	-	-	?
Erecta	<u>V.erecta</u>	50 (?)	32 (41a)	20 (24a)	63 (58a)	12 (17a)	37 (42a)	A

a Based on doubtful occurrence or on doubtful structures.

TABLE 5

Distribution of Groups of Alkaloids in *Vinca* Species[a]

Species	A-1.1	A-1.2	A-2.1	A-2.2	A-3	A-4.1	A-4.2	A-4.3	A-4.4	A-4.5	A-5	A-6	A-7	A-8	B-1	B-2.1	B-2.2	B-3.1	B-3.2	B-4	B-5.1	B-5.2	B-6
V. minor							(1)[b]	1						1	6	2	9				9	4	1
V. major	1		1				1		1		4										2		
V. difformis							3				2	1					2				1		
V. pubescens	1		1																				
V. herbacea	1	2	1	1	1								1				3						
V. libanotica																							
V. hausknechtii																							
V. erecta	3 (4)[b]		4				2 (3)[b]	1 (2)[b]	1	1			7		1	1	3 (5)[b]		2	(1)[b]	3 (4)[b]	1 (2)[b]	

[a] The figures correspond to the number of alkaloids belonging to a given group.
[b] Based on doubtful occurrence or on doubtful structure.

belonging to both classes A and B, usually with a predominance

of the former. The separation of V. minor from the other spe-

cies, and incorporation of it into a separate section (Table 2),

is also favored by the occurrence of certain typical bases a-

mong its alkaloids, particularly those of groups B-1 (vincamin-

orine family), B-2.2 (vincadifformine family) and B-5.1 (vin-

camine family). Similarly of significance, appears to be the

presence of several racemic alkaloids: minovine, vincanorine

[(±)-eburnamonine], (±)-ind-N-methylquebrachamine and (±)-

vincadifformine, even though the latter base is also present in

the alkaloid complex of V. difformis.

The second section - Major - consists of three species,

i.e. V. major, V. difformis and V. pubescens. The latter one

has been chemically studied only twice[19, 20] and the results ob-

tained in these studies are not very convincing. The number

of structurally clarified alkaloids (two !) is too low to allow

its serious sue for chemotaxonomic purposes. On the other

hand, it is not without interest, that the alkaloid carapanaubine

[= vinine (20)] (see Table 3) has not been found until now in any

other Vinca species, and as such could well be considered as

a typical constituent of this botanically unclearly defined species.

Concerning V. major and V. difformis, much more informa-

tion is available. Their alkaloid composition is very similar

and is even characterized by the co-occurrence of a number of
identical alkaloids (reserpinine, vincamajine, vincamedine)
(Table 3). This fact points to a close relationship of both of
the above-mentioned species, and allows their incorporation
into the same section. The evaluation of both taxa as independent species on botanical grounds, except for the opinion of
Pichon[1], is supported not only by the differences in amino acid
content of these species[21], but also by the occurrence of structurally different alkaloid types. While the incidence of a group
of alkaloids oxygenated in the 10 position (groups A-4.2, A-4.4
and A-5) (majoridine, vincamajoreine, akuammine, 10-methoxy-
vellosimine) seems to be a characteristic for V. major, the presence of 2-acylindoles, such as vincadiffine (group A-6), is
typical for V. difformis.[a]

Among three species forming the third section Herbacea,
only V. herbacea has been chemically studied. This species

_____

[a] We would like to point out that racemic vincadifformine
(Table 3), originally discovered in V. difformis, was also later
found in V. minor. But in the former species it was accompanied by its dextrorotatory form, and in the latter by its levorota-
tory form. This stereochemical characteristic must be considered of great chemotaxonomic significance.

has again a pronounced ability to synthesize a high percentage
of alkaloids of class A, most of which have a unique structure
with a hydrogenated heterocyclic E-ring (groups A-1.2 and
A-2.2). V. herbacea is also the only known source of this type
of alkaloid. This fact can therefore serve as a reliable criter-
ion for the incorporation of this species into a separate section.

Unfortunately, no chemical data concerning the character
of alkaloids in V. libanotica and V. haussknechtii are available.
Therefore, from a chemical point of view, it is impossible to
argue the appropriateness of placing them in the section Herbacea.

The last section Erecta (Vincopsis sensu Pobedimova)[2] is
monotypic and contains only the Middle Asian taxon V. erecta,
which is endemic to a limited area in the Fergan Valley (Uzbek
SSR.). In its alkaloid complex a large number of alkaloids
closely related to strychnine (group A-7, Table 3 and Table 5)
have been found. With the single exception of vincanine (= norfluorocurarine), which has also been discovered in the roots of
V. herbacea (Table 3), all of the alkaloids are characteristic for
V. erecta only. Therefore this complex can be safely used as
a significant chemotaxon for the species V. erecta.

In our opinion, the basic picture of the distribution of individual alkaloids and/or their groups discussed above, fits well

the taxonomy proposed by two of us (M. N. and F. S.). Addition-

al chemical and botanical information are necessary to prove

or correct the validity of this classification.

Our present chemical knowledge is not sufficient to add

constructively to the phylogenesis of the genus Vinca. The

main reason is our current limited knowledge of the detailed

biosynthetic pathways leading to the individual alkaloids, and

therefore we are not able to decide whether the alkaloids in

question were formed in the early, middle or late stages of

biosynthesis. If V. minor is really the closest taxon to the

original ancestor of the whole genus, as accepted presently,

the occurrence of alkaloids of the late stages of biosynthesis

(class B and C) would indicate phylogenetically older taxa,

while the incidence of alkaloid types derived from early stages

(class A) point to the phylogenetically younger taxa.

LITERATURE

1.   Pichon M. , Bull. Museum Hist. Nat. 23, 439 (1951).

2.   Pobedimova E. G. , Vinca L. in Flora SSSR. Vol. 18, p. 646;
     Izd. Akad. Nauk SSSR, Moscow (1952).

3.   Stearn W. T. , This Monography, p. 19.

4.   Paris R. and H. Moyse-Mignon, Compt. rend. 245, 1265
     (1957).

5.   Janot M.-M., J. Le Men and J. Garnier, Bull.Soc.Bot. France 1965, 118.

6.   Le Men J. and W.I. Taylor, Experientia 21, 508 (1965).

7.   Battersby A.R., Pure Appl. Chem. 14, 117 (1967).

8.   Pichon M., Mém.Mus.Hist.Nat. 27, 185, 204, 237 (1948).

9.   Nováček M. and Starý F., unpublished results.

10. Rakhimov D.S., M.R. Sharipov, N.Ch. Aripov, N.M. Malikov, T.T. Shakirov and S.Yu. Yunusov, Khim. Prirodn.Soedin. 1970, 713.

11. Rakhimov D.A., V.M. Malikov and S.Yu. Yunusov, Khim. Prirodn.Soedin. 1967, 310.

12. Abdurakhimova N., Sh.Z. Kasymov and S.Yu. Yunusov, Khim.Prirodn.Soedin. 1968, 135.

13. Il'yasova Ch.T., V.M. Malikov and S.Yu. Yunusov, Khim. Prirodn.Soedin. 1970, 717.

14. Rakhimov D.A., V.M. Malikov and S.Yu. Yunusov, Khim. Prirodn.Soedin. 1969, 461.

15. Rakhimov D.A., V.M. Malikov, M.R. Jagudaev and S.Yu. Yunusov, Khim.Prirodn.Soedin. 1970, 226.

16. Döpke W. and Meisel H., Pharmazie 26, 116 (1971).

17. Malikov, V.M. and S.Yu. Yunusov, Khim. Prirodn. Soedin. 1969, 65.

18. Hesse M., Indolalkaloide in Tabelle, Ergänzungswerk. Springer, Berlin-Heidelberg-New York 1968.

19. Orechov A., H. Gurewitch and S. Norkina, Arch. Pharm. 272, 70 (1934).

20. Abdurakhimova N., P.K. Yuldashev and S.Yu. Yunusov, Khim.Prirodn.Soedin. 1965, 224.

21. Paris R.R. and R.L. Girre, Compt. Rend.Ser. D, 268, 62 (1969).

CHAPTER 5

THE COMMERCIAL CULTIVATION OF VINCA MINOR

K. Szasz and G. Mark

Gedeon Richter, Ltd.
Budapest, Hungary

I. INTRODUCTION

The discovery of the biological activity and therapeutic effect of vincamine[1], the main alkaloid of Vinca minor, has necessitated the production of this agent on an industrial scale. After the habitats of this plant were located, and the industrial vincamine yield was found to be relatively low (0.0-0.4 g/kg dry material) the conclusion was drawn that the gathering of wild-growing plants was not feasible to meet the rawmaterial requirement of vincamine production on a larger scale. Because of this, commercial cultivation of this plant was initiated.

Distributed over wide ranges in the temperate zone, Vinca minor occurs almost solely as an undergrowth in oak-, beech- and hornbeam-forests. It is also fairly widely used as an ornamental plant, and is used for covering lawns in cemeteries

and parks. The plant stands may differ considerably in size.
There are places where V. minor covers an area of a few
square meters; but we also know of habitats where it spreads
continuously over several acres.

The natural vegetative circumstances of this plant might
suggest that it ought to be grown in shaded areas as an undergrowth. This idea is contradicted by our observation that the
growth rate of the plant is very low in the shade. This fact in
itself would preclude industrial cultivation of the native plant.
Further, to raise the plant as a forest undergrowth would
create unforeseeable agrotechnical difficulties.

Thus we felt that growing the plant for practical purposes
would be rewarding only if we are able to provide a natural
microclimate for the plant under field condition, and if we could
propagate the plant under such artificial circumstances.

To establish a plant culture one must have a ready source
of reproductive material. It is known that V. minor bears
small follicles which each contain 2-6 seeds. However the
fruits are very sparse, ripening is not uniform, and gathering
mature fruits is at best difficult. The plant does, on the other
hand, develope a root system with large numbers of runners,
and the stalks have a tendency to root. Because of the latter two

points, it became apparent that <u>V</u>. <u>minor</u> was a plant suitable
for vegetative propagation.

Since the observation of natural colonies of <u>V</u>. <u>minor</u> revealed considerable morphological differences, the question
presented itself as to whether a difference existed with respect
to alkaloid yield, and particularly vincamine yield, between
plants originating from different habitats.

To decide this question, material was collected from several hundred areas in Hungary and the samples were analyzed.
In this way it was shown that considerable differences existed
with respect to the total alkaloid yield and the vincamine content of <u>V</u>. <u>minor</u>. Szász and co-workers[2] reported that the total alkaloid and vincamine content in plants collected from
different places could vary by more than 50 per cent. Mathe
and Szabo[3] collected <u>V</u>. <u>minor</u> specimens from 87 localities
in Hungary - from forests, groves, parks, and cemeteries -
and determined the total alkaloid and vincamine content in the
leaves of dried samples. They found that the total alkaloids
varied from between 0.11 and 7.06 per cent, whereas the vincamine content varied between 0.02 and 1.75 per cent. Continuing
this work, Mathe and Precsenyi[4] found the vincamine content
of plant materials collected from various places in Hungary to
vary between 0.07 and 3.04 per cent.

This highly significant fluctuation in the quantity of vincamine in plants collected from their natural habitats necessitated
a consideration of the variations in field cultivated plants obtained from forests. To decide this question, Mark and Szász,[5]
in 1959, introduced material of significantly different vincamine-
yield, collected from different places, for field cultivation
under identical ecological circumstances, and studied the vincamine content of the crops by means of a preparative method
over a period of 3 years (Table I). These experiments and
studies allowed us to conclude that this shade-loving plant, forming undergrowths in forests, could be propagated and grown
successfully in sunny fields having a southern exposure, under
the conditions specified in the experimental part of this chapter;
and that the alkaloid content of such cultivated plants increases
considerably when compared with the alkaloids found in plants
growing in their original habitat. As concerns the increase in
alkaloids, the experiments have led to two interesting findings.

First, the trend of alkaloid production conforms to the
alkaloid content of the starting reproductive material, i. e.
alkaloid production following a starting material of relatively
low content will increase, but the yield will be low in relative
terms.

TABLE I

The Comparison of the Vincamine-Yield of the Feral
Population of <u>Vinca minor</u> L. and Those of Cultivated Plants

Finding Place	The Vincamine Yield of the Plant Collected from the Feral Population g/kg dried material V_1	The Vincamine Yield of the Cultivated Plant g/kg dried material V_2	$\dfrac{V_2}{V_1}$
I	0.73	1.68	2.30
II	0.38	1.12	2.95
III	0.42	0.68	1.61
IV	0.52	1.40	2.69
V	0.33	0.76	2.30
VI	0.34	0.62	1.83
VII	0.40	0.88	2.20
VIII	0.45	1.22	2.70
		Average yield:	2.30

Secondly, irrespective of the alkaloid content of the starting
material, the alkaloid content in plants grown under field conditions will at least be doubled, as compared with the content
in the starting material.

On the basis of the favorable results of small plot experiments, it was possible to scale up the cultivation on the industrial
scale, and; generally speaking, the conclusions, that we had
drawn proved to be correct. The experiments for industrial
cultivation were started on areas of 1-5 acres; some of the
plots were planted with reproductive material containing low
quantities of the agent; some plots with material of average
content, and others with material of relatively high content.
The plants from these experimental plots were examined every
week by preparative chemical analysis over 4 growth cycles.
The findings from these analyses warrant the following conclusions:

1.   The quantity of vincamine that can be produced from cultivated plants is on the average double the quantity from the same
plants grown under the original, wild, forest, conditions.

2.   Meteorological factors affecting the vincamine yield from
the plant grown under field conditions include the amount of
sunshine and the temperature, provided that the quantity of
water required for growth is available to the plant.

3.   A study of the four growth cycles showed that a spring and a late-summer peak appears in the total alkaloid and vincamine production of <u>V</u>. <u>minor</u> every year.   These peaks are present, with differences of one or two weeks, in April-May and in August-September, the shift in time is determined by the meteorological conditions of the particular year.

4.   The increased alkaloid accumulation occurs also under the circumstances of large-scale field cultivation; however, a certain negative trend was observed in the cultivated plants from these areas during the years of investigation, i.e., 1964, 1965, 1966, 1967.   On the basis of the findings of these four years, it was concluded that decreased vincamine production was attributed to certain biological processes arising from the conditions of the field cultivating over several years; yet such a conclusion is not supported by the findings of our work in 1968, which showed a slight increase in vincamine production compared to the preceding year as shown by Mark <u>et al.</u>[6]

In summary, it may be stated that <u>V</u>. <u>minor</u> can be propagated and grown in the field, out of its natural habitat, and that the crop yields vegetable matter whose industrial yield of vincamine is at least twice as high as that of vegetable matter taken from the best, wild, forest colonies.

A.   The Choice of the Cultivation Area.

In the course of growth experiments it was found that the
humid woodland habitat necessary for the growth of V. minor
could be adequately maintained in open areas, on arable land,
if sufficient moisture was provided by watering.

Consequently, the cultivation area should be a region which
is rich in precipitation, where the degree of humidity is high,
and where the climate is not extreme. The area should be protected from wind as far as possible, and should be near a
place where the plant occurs naturally, i.e. near oak-, horn-
beam- and/or beech-forests.

Vinca minor is demanding relative to soil requirements.
Thus, the soil should be rich in nutritive matter, should have
a pH between 6.5 and 7.5, should contain adequate calcium,
should be rich in vegetable mold, and should have a loose structure, but the water-holding capacity must also be high.

The first requisite for successful cultivation is to have an
adequate water supply available throughout the growing years,
and this fact should also be considered when the selection of
a site for the cultivation of V. minor is made.

Special care should be taken to select areas for cultivation
which are not heavily contaminated with weeds, especially
perennial weeds. The presence of such weeds requires much

additional work in the year following planting, and the process

of weeding also endangers those <u>V</u>. <u>minor</u> plants that are not

fully rooted. When eradicating weeds, considerable damage

may be caused to the stand if the trailers within the planting

spaces are exposed or cut during weeding. This may preclude

the natural propagation of the plant.

B.   The Reproductive Material

The shoots to be used for planting should be collected from

a native forest area from plants which have a vincamine con-

tent of greater than 0. 25 g/kg dry weight.

The shoots should be removed from the soil not earlier

than a few days prior to the anticipated planting. Each shoot

should consist of a sound root system and a superficial stalk

with 2-5 leaves. The shoots should be bundled in units of 200.

The bundles should be loaded on lorries in a root-to-root ar-

rangement, and in order to prevent heating they should be trans-

ported to the cultivation area, where the soil has been prepared,

as soon as possible, and preferably within 6 to 10 hours.

If necessary, the shoots can be stored in a pit for several

weeks. This is carried out by preparing a furrow about 20 cm

deep. The shoot-bundles are placed in the furrow with the

roots downward, the earth is placed on the side of the bundles

in a mound, pressed down firmly, and watered.

II.   AGROTECHNICS

A.   Preparation of the Soil

Since a culture of Vinca minor must be maintained and utilized for 6-10 years, special care should be devoted to the preparation of the soil before planting.

In addition to starting with a weedless soil, it is advisable to grow in the same area selected cereal crops, such as wheat or rye, as a green crop.

Adequate reserves of nutriment are also necessary for maintaining the culture for several years. One acre requires about 20 tons of manure and 0.2 tons of 40-per cent potash fertilizer, which should be spread on the stubble-field immediately after the harvesting of the green crop, and ploughed in to a depth of 25 cm.

It is advisable to carry out chemical weed control 2-3 weeks before planting, by applying 2 kg/acre of Actinit DF (4,6-bis-ethylamino-2-chlortriazine). The weed killer should be suspended in 400-600 litres of water when applied to the areas.

B.   Planting by Cuttings

The most suitable time for planting in central Europe is in late summer, at the end of August or the beginning of September. The advantage of autumn or late-summer planting are that the plants strike roots and grow strong before the frosts set in, and therefore survive the winter without excessive damage. On the other hand, a snowless, dry, very cold winter may cause frost damage. This can be overcome by planting in the early spring. If forced by circumstances, planting may be carried out early in spring, as soon as the field has dried up after the thawing of snow to such a degree that working is possible. If it is possible to complete planting at an early date, the growth of the spring culture may approximate that of the previous early autumn.

The shoots should be planted 30-40 cm apart in rows, thus 250,000-300,000 shoots are required for one acre. The roots of the shoots should be cut back to two-thirds of their size before planting.

Planting may be done by hand, using the dibble, or by a machine. Manual planting may be made in furrows 10-12 cm deep. The shoots should be placed in the furrows, the earth pressed against the roots at both sides. Care should be taken with either method that the shoots be placed deep enough beneath the soil surface, i.e. 20-30 mm of the stem above the roots should be in the soil. Planting is correct if it is not possible to pull out the shoot from the soil, holding them by the leaves, without ripping the leaves. Depending on the moisture

content of the soil, the cultivation areas should be watered with
a quantity corresponding to 20-30 mm rainfall, immediately
after planting.

C.   Maintenance of the Cultures

1.   Plant Nursing

Plant nursing should be started immediately after planting.
In case of autumn planting, measures for the winter protection
of the plantation should be taken before frost sets in, because
V. minor is somewhat frost sensitive. In its natural habitat,
V. minor is protected from frost by the falling leaves of shade
trees, and this protection must be provided also when the plants
are cultivated. The tender shoots may be covered with a blanket
of some suitable material - e. g. chaff - but this increases the
expense of the cultivation. It is therefore advisable to employ
the method of earth piles, commonly used in horticulture. The
earth blanket should be 6-8 cm thick.

As soon as the frost disappears in the spring, the plantation
should be rolled down without delay so that the earth loosened
by frost is compacted to the roots of the small plants. In order
to prevent the development of annual weeds, 1.5 kg/acre of
Actinit DT (4, 6-bis-ethylamino-2-chlortriazine), a selective
weed-killer, should be sprayed on the plantation after rolling.

It is most important during the first growth season to keep the culture free from weeds to the greatest degree possible, as well as to loosen the soil. Until the trailers of the plants extend to the spaces between rows, weeding within the spaces can be carried out by hoeing 4-6 times. However, along the planted rows, the weeds must be removed by hand. To keep the culture absolutely free from weeds is of extreme importance in the initial stage, because weed control at a later time is not possible without causing considerable damage to the stand. If optimum conditions for the growth of the culture are provided in the beginning, the plant will soon cover the entire area evenly, and will therefore act itself as a supressor of weeds.

2.   Watering

As has been mentioned previously, a prerequisite for the field cultivation of <u>V</u>. <u>minor</u> is to create the forest microclimate that prevails at the habitat of the plant. The moist, humid microclimate favorable and necessary for this undergrowth in forests can be provided in the open field by means of adequate watering. Consequently watering, equivalent to 400-500 mm of normal precipitation, must be provided in rainless periods during the growth season from April to October. Watering should be carried out by means of low-intensity sprink-

ling equipment with spray heads in such a number that a quantity of water corresponding to 20-40 mm of rainfall should be
applied on each occasion. It is advisable to carry out watering
at sunset or at dawn.

3.   Chemical Weed Control

Vinca minor, fortunately, is relatively resistent to chemical agents. For this reason it is possible to use many different
weedicides as shown by Svab and Foldesi.[7]

a.   Weed Control in New Cultures

Chemical weed control is preferably carried out by the
application of 2 kg/acre of Actinit DT. The chemical should
be applied in 250-300 litres of water per acre. Any type of
powder sprayer is adequate. If the plant culture is on a slope,
special care should be taken that the spray be distributed evenly.

The best time for spraying is in November, immediately
after the autumn plowing. It is our experience that the weed-
killing action of Actinit DT, chiefly as it concerns couch grass
(Agropyron repens), is more efficient at this time of the year
than if spraying is done in the spring.

If there is no possibility for autumn spraying, this work can be performed at any time in the winter or in the frost-free months of early spring.   The only precondition is that the soil should not covered by snow, and that the spray should be applied directly to the soil.   The Actinit DT solution applied to the solid soil becomes fixed at once and cannot be washed away. It can be sprayed on frozen soil without involving any loss. Spraying in the spring should be timed in such a way that it is completed about 10 days before planting.

Any type of superficial soil cultivation can be carried out in areas treated with chemicals.   If chemical treatment is over-due, it should not be applied before the shoots have taken root. During the first week of the growth season, following application of Actinit DT, the leaves of <u>V</u>. <u>minor</u> become yellow with spots, but this not adversely affect the development of the plant.

b.   Chemical Weed Control of Several Years Old Plants.

Chemical weed control of such plants should be carried out in a similar manner.   Depending on the degree of weed con-tamination of the culture and on the hardness of the ground, the dosage of Actinit DT should be 2-3 kg/acre, and is to be applied preferably in autumn, or in very early spring before vegetation sets in.   If spraying is carried out in the spring, it is advisable

to do so immediately after the snow disappears, when the surface of the soil, especially in the early hours, is frozen and
can hold the weight of the spray machine. Vinca minor sprouts
and flowers very early. A flowering stand must not be treated
with chemicals. If weedicides must be applied during the growing season, they should be applied only after flowering is complete.

c.   Repeated Application of Weedicides

Vinca minor tolerates chemical weed control by Actinit DT
so well that treatment can be repeated over a period of several
years. In determining the dosage, the degree of weed pollution
of the area must be taken into account. In cases of intensely
weedy soil, the Actinit DT dosage should be 2 kg/acre in the
first year, and 2-3 kg/acre in subsequent years. If the area
still contains many weeds amounts of 6 kg/acre of Actinit DT
can be applied, but an interval of at least a year must be observed. In such a cases Actinit DT accumulates in the soil to such
a degree that repeated spraying does not lead to any further
weedicide action, but rather, it may reduce the growth of the
plant.

d. The Weedicide Effect of Actinit DT (4, 6-bisethylamino-2-
chlortriazine)

Actinit DT is generally known as a herbicide having a broad
spectrum, which is not decomposed completely during the first
year of application, and which continues to act during the subsequent year following application. It kills most annual, dico-
tyledonous weeds, and also has a good effect (ca. 80%) on annual monocotyledons.

There are several species among the deep-rooting annual
weeds generally known for their resistance to Actinit DT.
Similarly, a considerable resistance is displayed by several
species of the perennial monocotyledonous couch grasses (see
Table 2).

Since the occurrence of the above species of weeds may be
expected even following repeated applications of Actinit DT, it
is advisable, especially in the first year of culture, to supple-
ment chemical weed control with mechanical and manual weeding.
Freedom from weeds can be insured during subsequent years by
following this procedure.

Analytical investigations carried out simultaneously showed
that chemical treatment does not adversely affect the alkaloid
and vincamine production and content of <u>V</u>. <u>minor</u>.

TABLE 2.

Weeds Resistant to Actinit DT

Resistant	Moderately resistant
Bindweed (Convolvulus arvensis)	Canadian thistle (Cirsium arvense)
Horsetails (Equisetum species)	Wheath grass (Agropyron repens)
Silver-weed (Potentilla species)	Common comfrey (Symphytum officinale)
Wild pea (Lathyrus tuberosus)	Knotweed (Polygonum species)
Dewberry (Rubus caesius)	
Wire grass (Cynodon dactylon)	
Johnson grass (Sorghum halapense)	
Colts foot (Tussilago farfara)	

4. Nutriment Supply

The defoliation of deciduous trees taking place annually in

the natural habitat of V. minor provides not only an excellent

frost-protective winter blanket, but insures the availability of

a vegetable mould rich in organic substances. It follows from

this that <u>V</u>. <u>minor</u> prefers a soil that is abundant in these nu-
triments. As has been indicated previously, the perennial
nature of this plant requires an increased supply of nutritive
matter.

Therefore, in addition to insuring adequate nutriment when
the soil is being prepared prior to planting, a continous and
regular supply of nutritive matter must be applied while the
culture is maintained.

The culture should receive top-dressing twice a year as a
rule. The first should be applied in early spring and the second
in June. A fertilizer corresponding to a N content of 20-25 per
cent, and consisting of ammonium nitrate and calcium, should
be spread over the area, the quantity being 70-100 kg/acre on
each occasion. If the soil is calciferous, it is advisable to ap-
ply ammonium sulfate instead of ammonium nitrate.

The fertilizer should always be spread before weeding.
Hoeing and spreading should be followed by watering at all times.
When the area is completely covered by the <u>V</u>. <u>minor</u>, hoeing
is not possible.

5.   Parasites and Protection

Up to now <u>V</u>. <u>minor</u> has not been grown in the field on a
large scale, therefore our knowledge relating to the diseases

and parasites of this plant is incomplete. It is our experience
that plants occurring in their native environment apparantly
thrive well, and that parasites or diseases are seldom encountered. However, as soon as a feral plant is used for cultivation,
damage due to parasites or diseases caused by unknown patho-
gens, are encountered.

Considerable damage can be caused in new plantations by
the grubs of various beetles, i. e. maybeetle species (Melolon-
thinae), mole crickets (Gryllotalpa), and by parasites in general that tend to consume the tender young plants.

The most efficient protection against such parasites is
thorough soil disinfection before planting. In order to prevent
damage by grubs it is advisable prior to planting, to immerse
the cut roots of the shoots in clay pulp that contains 1 per cent
Aldrin (hexachlor-hexahydro-dimethanonaphtalene). Because
this operation increases the labor involved in mechanized planting, the problem may be solved by spreading at the time of
soil preparation a superphosphate fertilizer that contains the
required amount of Aldrin.

A spotty decay may appear in cultures of V. minor late in
the spring or early in the summer, when the upright tender
sprouts dominate. In our work, a superficial survey indicated
yellow spots in a field of V. minor. Upon a more thorough ex-

amination of the withered plants it appeared that the color of

the diseased sprouts became darker downward, turning brown,

then totally black. The Base of the blackened stalks were moist

to the touch and could be torn off easily from the root. In case

of a massive infection of the plant, even the roots were black

and evidently rotten. By cultivating the pathogenic agent, and

by means of artificial infection, Kaszonyi[8] found that the disease

was caused by the fungus <u>Phoma</u> <u>vicae</u> <u>minoris</u> Kaszonyi. Dam-

age due to this fungus depends on meteorological factors. A

cool, humid, atmosphere increases damage, while damage is

less in dry, sunny weather. The disease may involve as much

as 25-40 per cent of the stocks; but it has observed that the

number of infected stocks decreases as summer advanced.

Infection with <u>Phoma</u> can be prevented by spraying with

a suitable fungicide.

D.   <u>Harvesting</u>

1.   Cutting

A culture planted late in summer or in autumn can be har-

vested after about two years. A culture planted in the spring

may at times become so vigorous that by autumn of the follow-

ing year, the crop can be gathered. Cutting of the crop should

be carried out preferably late in August, but not later than by
the middle of September.

Since the purpose of cultivating V. minor is to obtain
plant material for the industrial production of vincamine, the
time of harvesting should be adjusted to the period when the
dry vegetable matter to be obtained has the highest vincamine
content that can be extracted on an industrial scale. A spring
and an autumn peak in alkaloid content are evident during the
course of the growth season. However, the moisture content
and consistency of these plants necessitates that they be cut in
the autumn. In the autumn the sprouts and the stalks lie flat on
the surface of the soil, therefore cutting is only possible with
a knife, or, at best, with a sickle. In the spring, on the other
hand, during the period of flowering, the tender sprouts of high
moisture content stand upright and permit cutting by means of
a scythe, and yield a drug of high alkaloid content.

If the harvest is planned for autumn, the culture should
be given top-dressing with a nitrogen-containing fertilizer, in
addition to the routine nursing of the plants. This fertilizer
should be applied twice, 4 and 2 months before the anticipated
harvest. Fertilization should, of course, be followed by watering.

In the second year of culture, harvesting should be at the end of August or at the beginning of September. The day of cutting should be fixed according to the trend of the vincamine content of samples investigated by trial cuttings every week. In this way, the maximum alkaloid content the shoots is known; thus the changes in alkaloid content can be forecast from the trend of the curve of alkaloid content. The cutting of the crop should be preceded by 5-10 days of bright, warm weather if possible.

It is important for the regeneration of the culture that 3-5 cm long stumps of stalks be left after cutting. The cut vegetable matter contains 80-90 per cent water in the spring and 60-70 per cent in the autumn. Due to the considerable difference in water content, the harvested plants require different handling.

2.   Drying

The spring material, of high water content, is extremely difficult to dry. Consequently, cutting should be started only if facilities for artificial drying are available. Plants should be dried within 1-3 hours following the harvest.

The autumn crop of lower water content is not so sensitive, but drying must be started within half a day following harvest. If surfaces of sufficient size are available, the autumn crop

may be dried naturally, in 5-10 cm thick layers, placed in build-
ings having good ventilation.

The vincamine in V. minor is relatively insensitive to the
temperature of drying. Temperatures of up to 150° C do not
decompose the vincamine content of the dry drug. Thus, it is
possible to dry the plant material quickly by artificial heat.

The continuous, countercurrent HBM drier (Hans Binder
Maschinenfabrik, Freising-Marzling, GFR), with temperature
regulation, has proved to be the best for drying V. minor.
Owing to their high capacity, Cyclone driers, such as are used
for alfalfa meal, can also be used, but then it is rather difficult to regulate the final moisture content of the product.

REFERENCES

1. L. Szporny and K. Szasz, Arch. Exptl. Pathol. Pharma-
    kol, 236, 296 (1958).

2. K. Szasz, T.M. Kovats, E. Karacsony, Cs. Lorincz and
    J. Bayer, Planta Med., 7, 234 (1959).

3. I. Mathe and Z. Szabo, Herba Hung., 2, 289 (1963).

4. I. Mathe and I. Precsenyi, Acta Agron. Acad. Sci. Hung.,
    15, 274 (1966).

5. G. Mark and K. Szasz, I. Hungarian Symposium on Medi-
    cinal Plants, 1959, Pecs. (Unpublished data).

6. G. Mark, L. Gracza and K. Szasz, Herba Hung., 8, In
    press. (1969).

7.   J. Svab and D. Foldesi, Herba Hung. <u>5</u>, 254 (1966).

8.   S. Kaszonyi, XIV. Hungarian Scientific Conference on
     Plant Protection, Budapest. 1964. (Unpublished data).

CHAPTER 6

THE PHARMACOLOGY OF
VINCA SPECIES AND THEIR ALKALOIDS

Milos Hava

Department of Pharmacology
University of Kansas Medical Center
Kansas City, Kansas

I. INTRODUCTION

In spite of the large number of alkaloids obtained from the

six Vinca species under discussion, only a few alkaloids have

been pharmacologically screened and even fewer have been clin-

ically tested. No well documented pharmacological literature

is available concerning the pharmacology of the two possible

subspecies[1] of Vinca major, V. difformis and V. pubescens (ex-

cept an old paper of Orechoff et al.[2]). The use of the different

Vinca species in folklore medicine was described in chapter 2.

The greatest part of the present literature is concerned with

vincamine from V. minor, which is an important alkaloid of

the majority of other species too.

*Present address: Peoria School of Medicine, University of

Illinois at the Medical Center, Peoria, Illinois 61606.

TABLE 1

Species	Alkaloids pharmacologically tested	Pharmacological effects	LD$_{50}$ in mice mg/kg	Clinically tested	Comments
Vinca minor	total alkaloids	hypotensive, sedative,	i.v. 24 i.p. 76 p.o. 500	as hypotensive	Russian VIPAN
		spasmolytic, positive curare-like, inotropic, gangliolytic, negative chronotropic, local anesthetic,			
	vincamine	hypotensive, sedative, spasmolytic, hypoglycemic bioamines release	i.v. 75 s.c. 1.0	as hypotensive	Hungarian DEVINCAN
Vinca major	total alkaloids	hypotensive spasmolytic, gangliolytic, local anesthetic	i.v. 37		

Species	Alkaloid	Action	Dose
<u>Vinca</u> <u>herbacea</u>	perivincine mixture of vincamine + vincine	brief hypotensive	
	total alkaloids (vincamine + vincanine present)	curare-like, gangliolytic, hypotensive	s.c. 238
<u>Vinca</u> <u>erecta</u>	total alkaloids	analeptic, sedative, oxytocic	i.v. *12.9 as analeptic
	vincamine	viz. <u>Vinca minor</u>	
	vincanine	strychnine-like, papaverine-like	i.v. * 7.5 as analeptic
	vincanidine	adrenolytic	
	ervinine (kopsinine)	strychnine-like, analeptic	s.c. 125

pseudo-kopsinine	strychnine-like, weak analeptic	s.c. 60
tombozine	sedative, hypotensive, ganglioblocking, heart depressive, papaverine-like	i.v. 65 s.c.325
vineridine	sedative, spasmolytic, oxytocic	i.v.125 i.p.485
vinervine + vinervinine	hypotensive	i.v. 25 i.p.100
vincarine	sedative	i.p.330 s.c.520
ervamidine	sedative, hypotensive, muscle relaxant	i.p.391 s.c.550

| ervamine | sedative, oxytocic, hemostatic | i.v. 90 i.p.200 p.o.370 |
| vinerine | sedative | |

* in the rabbit

From Table 1 it is evident that Vinca species differ quite
substantially in pharmacological properties from their total
alkaloids. Hypotensive effects have been reported for all of
them, which may be due to vincamine and vincamine-like alkaloids, plus reserpine and its derivatives, sedative in V. minor
and V. erecta, spasmolytic in V. minor and V. major, curare-
like and gangliolytic in V. herbacea and V. minor, analeptic
and oxytocic in V. erecta. Probable explanation of these overlapping activities is that the same alkaloids are found in several species (see vincamine, vincanine, etc.).

Clinical interest has been directed mainly to the effects
on circulation (hypotension, cerebral circulation), and partly
to the central effects (analeptic, sedative). The only pure alkaloid therapeutically used in Eastern Europe is vincamine (Hungarian preparation Devincan), though no real evidence of its
advantages against the standard treatment of hypertension
were given and its effects were at best mild.

There is no doubt that the pharmacological effects of
Vinca alkaloids thus far described are interesting, and the
high percentage of alkaloids not yet pharmacologically evaluated gives a reasonable promise that therapeutically useful
properties will be found as shown in the example of alkaloids

from Catharanthus species (vincaleukoblastine, etc.).

II.   THE PHARMACOLOGY OF VINCA MINOR

Quevauviller et al.[3,4] first tested the total alkaloids from
V. minor in 1954. They found the p. o. LD_{50} to be 500 mg/kg
in mice and characterized the action as gangliolytic and spasmolytic. The intraperitoneal LD_{50} in mice was 76 mg/kg, i. v.
LD_{50} = 24 mg/kg; in slow i. v. infusion the minimal lethal dose
in guinea pigs was 16 $\pm$ 1. 9 mg/kg. A hypotensive effect per-
sisting for 10-30 minutes was evoked in dogs under chloralose
anesthesia by doses of 5-10 mg. The hypotension was not
blocked by atropine, and only partly blocked by nicotine. Direct spasmolytic effects on vascular and intestinal smooth
muscle were observed. A 1% solution of alkaloids had a local
anesthetic effect corresponding to 0. 7% cocaine. Slight stimulation of respiration and negative inotropic effects were noticed.
Later Gazet and Strasky[5] described a coronary dilation and a
negative inotropic effect by galenic preparations of V. minor.

In 1956 Polish authors[6-8] tested the whole dry plant and
observed a blockade of the sympathetic ganglia, stimulation of
the parasympathetic system, central depression of the vaso-

motor system, and decreased sensitivity of the peripheral vaso-
motor system. The doses used were 0.025-0.3 g/kg. Different animal species (dogs, rabbits, cats, rats, and mice) were
tested. The decrease of arterial blood pressure was the smal-
lest in rabbits and the greatest in dogs. It was of the longest
duration in cats. A single-intravenous dose which decreased
the blood pressure was not influenced by the cutting of the vagus
nerve. Atropine made the decrease somewhat smaller; the
antihistamine Antistine^R had no effect on it. Decerebration
prolonged the hypotensive action in the dog, and abolished it
completely in the cat. The hypertensive carotid reflex was
decreased. No effect on adrenaline, ephedrine or vasopressin
hypertension was observed. Acetylcholine, prostigmine and
dihydroergotamine hypotension was not influenced. Effects of
V. minor were not altered by Pendiomide^R, but were prolonged
by procaine.

The heart was slowed in the dog and cat, and the vessels
of the skin and muscles were dilated. Atropine slightly weakened those reactions. The contraction of the nicitating mem-
brane in the cat, after electrical stimulation of preganglionic
fibers of the cervical sympathetic trunk, was abolished.

Small concentrations were without effect on isolated vascular preparations of the dog, rat, and frog, while high concentrations caused a contraction. No effect, even in great concentrations, was found in the isolated frog heart. Small doses evoked a positive inotropic effect in the isolated heart of the cat and the rabbit; large doses had a negative chronotropic effect.

The tension and peristalsis of the small intestine of the cat and rabbit in situ, as well as of the isolated intestine, were increased. Adrenaline weakened this effect, and atropine and papaverine blocked it.

No local irritation was observed. A hypoglycemic effect lasting three hours was evoked after intravenous injection of 0.05 g/kg. Weak sedative effect was found after i.v. injection of 0.5 g/kg. The LD_{50} in mice of the dry powder was 1.4 g/kg i.p.

Voskoboinik[9] described the properties of the Russian preparation Vipan containing total V. minor alkaloids. A dose of 10 mg/kg caused a prolonged lowering of blood pressure in cats and dogs. Bulgarian authors[10], studying the total alkaloids from V. minor, described a curare-like effect of the alkaloids in rabbits, cats and rats in doses of 6-7.5 mg/kg. Lowering of blood pressure was seen after doses of 1-3 mg/kg. The authors presumed the presence of two different groups of alka-

loids; one with mainly hypotensive effects, and the second with

curare-like properties. Favorable effects were described in

patients with hypertension. A full clinical report was first pub-

lished in 1957 by Szczeklik[11]. Thirty persons with arterial hy-

pertension, in different stages of the disease, were submitted

to treatment with powdered herb of V. minor in doses of 3.0 gm

per day orally. Systolic and diastolic pressure decreased by

20-30 mm Hg in 10 patients out of 12 in the I and II stages of the

hypertensive vascular disease, and in eight patients out of 10 in

later stages of the hypertensive vascular disease. Combined

treatment with Largactil[R] administered orally in daily doses of

75 mg, and V. minor herb in doses of 3.0 gm per day was given

patients with threatening cerebral hemorrhage. Out of eight

patients thus treated, four showed considerable improvement.

No side effects were noticed. The results obtained with V.

minor were about the same as the results obtained with Serpa-

sil[R], barbiturates and purines. In the early stages of the hy-

pertensive vascular disease V. minor gave even better results

than the treatment with barbiturates and purines. A tincture

of V. minor was also clinically tested in the USSR by Shelekov

[12,13]. Ninety patients with different stages of hypertension

were given a 5% tincture of Vinca minor orally at doses of 1-
1.5 tablespoonfuls three times daily for 10-20 days. A hypotensive effect was observed in 74 of the patients. The effect
was most marked in the first and second stages of the disease.
The hypotensive effect was accompanied by a sedative and a
spasmolytic action. In most of the patients it also increased
diuresis. The tincture seemed to be well tolerated and no
side effects were noted.

Although it has been indicated that vincamine and vincamine-like alkaloids have a principle role in the pharmacological and clinical effects of the above described preparations,
the other alkaloids present definitely modify the effects. Favorable clinical effects were reported, but the experimental
studies were done on a small number of patients without good
controls and double-blind evaluation, and have only an arbitrary
value. Nevertheless they were promising enough to start a
serious pursuit of the effective principle-alkaloid vincamine.
As the latter part of this chapter will show, the interesting
species differences found by Polish authors were seen with the
pure alkaloid as well.

III. THE PHARMACOLOGY OF VINCAMINE

In 1954 Raymond-Hamet[14] was the first to study the pharmacological properties of vincamine. In dogs, after doses of

2-6 mg/kg i. v., he observed a short-lasting hypotension, sympathicolytic effect and acetylcholine-like effect on the heart
and respiration.

Intensive studies were undertaken later by Hungarian workers[15,16,17]. Sedative effects in mice and rats, increasing
with the dose, were seen after doses of 2.5-10 mg/kg. The effect of thiobarbital was prolonged by the same dose; metrazol
seizures were not affected. Bi-phasic lowering of blood pressure was described in rabbits after 5 mg/kg. The first phase
was only short-lasting and deep, the other reached a maximum
in two hours, and lowered the blood pressure by 18-48%. In
cats neither ganglioblocking nor adrenolytic effects were observed. Also, sedation was observed in rabbits, dogs, and
cats. The intravenous LD_{50} in mice was 75 mg/kg; in rabbits
33 mg/kg. The subcutaneous LD_{50} in mice was more than
1,000 mg/kg. In chronic experiments in rats no diarrhea was
observed, in contrast to reserpine.

The Czech author Krejci[18] was not able to reproduce most
of these results and after doses of 1-5 mg/kg did not find any
secondary lowering of the blood pressure in cats. These
doses evoked a cardiac depression and depression of respiration. No sedation or prolongation of anesthesia after barbiturates was found in mice. When much lower doses (0.02-0.08

mg/kg) were used, a prolonged hypotension in the cat was observed, without changes in heart action and respiration, and a slight ganglioblocking effect was described. The best hypotensive results were found in hypertensive rats.

The Hungarians very soon thereafter tried vincamine (as Devincan[R]) in the clinic[19, 20, 21], using 5-20 mg/patient (p. o.) daily for three to eight months, which is closer to the doses used by Krejci. No side effects were seen even after a single dose of 100 mg per patient and the therapy was effective in 16 of 31 patients with hypertension of different origin[5].

In rats, vincamine caused a hypoglycemia at a dose of 0.2 mg/kg and prevented hyperglycemia after dextrose, but not after adrenaline[22].

Machova[23], in contrast to the findings of Hungarian authors[22], was not able to reproduce the results of the latter group blood sugar levels using rabbits instead of rats. A dose of 2.0 mg/kg caused hyperglycemia, with a subsequent increase in insulin level, and secondary hypoglycemia.

Gorog and Szporny[24, 25] using high doses in rats (50 mg/kg i. p.), found a reduced noradrenaline and serotonin content in the brain, intestines and suprarenals. This subtoxic dose resembled the effect of reserpine, but was too high to help elucidate the mechanism of vincamine action. Linet et al.[26] later published their findings of an increased elimination of 5-HIAA

(hydroxyindole acetic acid) after smaller doses of vincamine

(5 mg/kg i.p.) in rats by 32%. Diuresis was diminished by

27%. In six volunteers, after a dose of 4 mg/person, these

results on 5-HIAA excretion were confirmed. The effect lasted

8-10 hours and the increase was 16-164%.

Molnár and Szporny[27] described the effects of vincamine

on intestinal smooth muscle in different species (colon and ile-

um of the rat, rabbit and guinea pig). At 10^{-6} g/ml a contrac-

tion (except guinea-pig colon) as a direct effect on smooth musc-

le was described; at a higher concentration (10^{-4} g/ml), a re-

laxation and spasmolytic effect were found in the guinea pig and

rabbit. Both effects were reversible.

Machová and Mokrý[20] gave another explanation for these

effects and expressed doubts about the direct effect on the

smooth muscle. Guinea pig ileum contractions were observed

proportionally at doses of 0.2 to 13 μg/ml. At doses of 25-400

μg/ml a relaxation was found. The contraction was completely

abolished after morphine (0.1 μg/ml). The ratio of doses caus-

ing equal contractions in the presence and absence of atropine

(1 μg/ml), was on the average 10. On the isolated rabbit ileum

the vincamine contraction was not inhibited by atropine. Hex-

amethonium (5-80 μg/ml) either did not effect the vincamine

contraction or it partly depressed it. The vincamine contraction could be inhibited by a depolarizing type ganglionic block-
ing agent (nicotine, 5 μg/ml). The relaxation caused by high
doses of vincamine was inhibited by guanethidine (2 μg/ml), and
a motor response was revealed. On the basis of these experiments they concluded that both the excitatory and inhibitory
actions of vincamine were due mainly to neuronal stimulation.

Machova[29] also observed a contraction of isolated tracheal
smooth muscle which was directly proportional to the log dose
after vincamine in concentrations of 1.4-110 μM. A latency
period and tachyphylaxis were observed. Rabbit trachea, in
contrast to the trachea from dogs, guinea pigs, cats and sheep,
was not responsive. Vincamine potentiated the effects of acetylcholine, histamine, 5-HT, and was not influenced by atropine, morphine, hemicholine, 2-phenylbenzylaminomethylimi-
dazoline, LSD and calcium acetylsalicylate. After premedica-
tion with acetylcholine, the latency of vincamine contractions
was shortened from 767 ± 188 sec to 37 ± 6 sec. Histamine
and 5-HT gave similar effects. Reserpine and guanethidine
potentiated the contraction produced by vincamine.

The CNS effects were also studied by Machova and Selecky[30]. In mice, when doses of 5-10 mg/kg were administered i.
p., decreased motility lasting 20-50 minutes, and potentiation

of thiopental anesthesia were found. Quarternarized vincamine
reduced the response of neuromuscular preparations of cat
gastrocnemius muscle; vincamine did the same only in suble-
thal doses. The carotid-occlusive reflex in cats was not influenced at all (5 mg/kg i. v.).

Kurmukov[31-34] studied the central effects of i. v. and i.p.
administrated vincamine on the rabbit encephalogram. In
doses lower than 2. 5 mg/kg an activation of biopotential was
observed, followed by delayed high-amplitude rhythm, specific
to tranquilization of the central nervous system. In higher
doses, reaction of brain evoked by contralateral optic stimulation was inhibited.

Polish authors[35], working with vincamine and isovincamine, found no effect on cat blood pressure after 0. 3 mg/kg
i. v. and a decrease of blood pressure in dogs after 2. 5 mg/kg,
with depression of respiration. Depression of the heart was
also observed.

Kovách, in 1967[36], compared the pharmacological effects
of vincamine with four synthetic derivatives. Circulation and
respiration in the dog were followed after i. v. administration
of 2. 5 mg/kg of the drugs. The results indicated that the hypotensive effect of vincamine was due mainly to a reduction of
peripheral vascular resistance; diminution of cardiac output

and depression of respiration may also have had some effect
in the response.

Experiments with hexahydrocanthinone showed that a change
in the ring structure of vincamine eliminates the circulatory
effects of the compound while its other properties, such as the
stimulation of respiratory rate, induction of seizures, arousal
reaction, etc., became more pronounced.

Changing the ring structure of substituents of vincamine
resulted in apovincamine; haemodynamic properties were essentially identical with those of vincamine.

The compounds N-methyl-vincamonium iodide and chloride,
which are the N-quaternary derivatives of vincamine, caused
a drop in arterial blood pressure similar to that elicited by
vincamine; however, their effect was not associated with any
decrease in cardiac output or with apnoea.

In dogs Szabo et al.[37] observed a brief effect on renal he-
modynamics following the pattern of blood pressure and circulation changes.

The resorption, elimination and distribution of vincamine
in the rat was studied with ^{14}C labelled vincamine[30]. Resorption from the gastrointestinal tract was rapid. Vincamine was
quickly and evenly distributed unchanged throughout the body.
Elimination was rapid, and metabolites as well as unchanged

vincamine were found in the urine mainly in the first 24 hours.

Szporny and Gorog[37] stressed the low toxicity, mild prolonged action without sedation, and lack of side effects on the
gastrointestinal tract observed in patients. Their explanation
of its pharmacological action was 1) short stimulation of the
cholinergic system causing a brief drop in blood pressure, followed by 2) noradrenaline liberation, causing a brief increase
in blood pressure, and as a consequence of the catecholamine
depletion 3) clinically observed long-lasting hypotension.

The clinical report of Hargitai and Werkner[40] showed 58%
success in 78 patients (ages 16-78 yrs, 49 men and 29 women)
with different forms of hypertension after three to six days,
and following three weeks. These results must be considered
more of an observation than a real clinical trial.

Solti et al.[41], on the basis of their experiments on dogs[42],
studied the effect of Devincan[R] on cerebral circulation and
oxygen consumption in 11 patients. All patients were hypertensive. Twenty minutes after 5 mg i.m., a slight lowering
of blood pressure was observed, cerebral blood flow increased,
vascular resistance decreased, an oxygen consumption was
not changed. Devincan[R] proved to be useful in cerebral angio-
spastic syndromes[43, 44].

TABLE 2

Pharmacological Effects of Vincamine in Animal Experiments
Doses 0.02 - 10.0 mg/kg

Pharmacological effect	Authors					Comments
	Hungarian	Polish	French	Czechoslovakian	Russian	
Hypotension	++	±	±	+ in low doses	+	main clinical effect
Effects on cerebral circulation	+	+	+			clinical effect
Effects on renal circulation	+					
Heart depression	±	+	±	±		---
Respiration depression	±	+	+	±		---
Smooth muscle relaxation or contraction	+ direct			+ indirect		
Central sedative	+			±	+	observed in human too
Release of bioamines	+			+		observed in human too
Hypoglycemia	+ rat			rabbit hyperglycemia secondary hypoglycemia		

Ravina[45] summarized the therapeutic reports on vincamine
and stressed the favorable effects on cerebral circulation.

Summarizing the work done with vincamine (see Table 2),
there are several contradictions in the findings of different
authors, which can be only partly explained by species and me-
thodological differences. The effects of vincamine are multiple,
and are very much dependent on animal species and doses.
Most of these effects were not observed in clinical observations
where comparatively very low doses were used, e. g. 5-10-30
μg/day/patient = 0.1 - 0.6 mg/kg (against 2-10 mg/kg animals).

The effects on circulation were found therapeutically useful in the clinic. No really reliable clinical trial has been documented in the literature, and most of the reports can be described as clinical observations without proof of superiority of
vincamine treatment over presently employed treatments for
the lighter forms of hypertension.

IV. PHARMACOLOGY OF VINCA MAJOR

Sixteen alkaloids are known from this species. Orechoff[2]
reported a prolonged hypotension in experimental animals with
10 mg/kg i. v. of total alkaloids and with vinine, (synonymous
with carapanaubine) and pubescine (synonymous with reserpine).

The pharmacology of the total alkaloid extract of <u>V</u>. <u>major</u> was studied by Quevauviller[46, 47]. The extract was quite toxic: intravenous LD_{50} in mice was 37 mg/kg; showing sympathycolitic and parasympathicolytic properties in experiments on dogs in doses of 5-10 mg/kg i.v. This dose lowered blood pressure and caused peripheral and coronary vasodilation. Intravenous injection of the total alkaloids in dogs under chloralose anesthesia increased slightly the frequency of respiration, without changing the amplitude, and caused a hypotension which increased with the dose. This hypotension was not influenced by atropine, but was blocked by nicotine. Central sympathicolytic effects, gangliolytic, and spasmolytic effects were observed. No direct effect on the cardiac muscle was found.

The spasmolytic action tested on intestinal smooth muscle (rat duodenum), at a concentration of 10 μg/ml, was 1,000 times less than that of atropine. Local anesthetic action on the rabbit cornea was similar to the action of <u>V</u>. <u>minor</u> total alkaloids. When compared with the alkaloids from <u>V</u>. <u>minor</u> more peripheral action and less central action was found.

Farnsworth[48] isolated and studied the properties of the alkaloid perivincine. It represents 1.6-1.8% of total alkaloids and was proved later to be a mixture of vincamine and vincine. When hyper- and normotensive rats were treated with doses of

5-10 mg/kg i. v. , a blood pressure lowering by 27-40% was observed which lasted only 20-30 minutes.

V.   PHARMACOLOGY OF VINCA HERBACEA

The total alkaloid extract was pharmacologically tested by Bulgarian authors[49-51]. Among 14 alkaloids described from V. herbacea are vincanine and vincamine with well defined pharmacological properties. The most interesting effect was the curariform action observed after i. v. administration of 8 mg/kg of the total alkaloid extract to rabbits and cats. The blockade of neuromuscular synapse in cats and the positive head drop test in rabbits after doses from 8-40 mg/kg were only partly blocked by prostigmine and other cholinesterase blockers (eserine, nivaline). A dual character of the neuromuscular blockade was suggested: depolarization and competition. Smaller concentrations (2×10^{-5}) did not produce contraction of the m. rectus abdominis of the frog, but increased the sensitivity to a second equal dose of total alkaloids. With the second dose, a strong and longlasting curariform effect was observed. Neither the contraction produced by two sequential small doses, nor the contraction evoked by a large concentration (2×10^{-4} to 1×10^{-3}), were antagonized by competitive muscle relaxants (curare). In later experiments on cats, chickens and frogs the dual character of the muscular blockade was confirmed with the competitive mechanism prevailing.

Depending on the dose injected (0.1-20.0 mg/kg), the i.v. administration had a hypotensive effect lasting 15-20 minutes (10-40% lowering of blood pressure) in rabbits and cats. A complex mechanism of hypotensive action was proposed: depression of the central nervous system, gangliolytic effect and peripheral effect on the blood vessels. The effects on cardiac activity were considered not essential. Small doses stimulated respiration, large doses caused a respiratory paralysis (depression of the respiratory center and curare-like effects). Ganglioblocking was more pronounced on parasympathetic ganglia than on sympathetic ganglia. Potentiation of an alpha effect and blockade of beta adrenergic effects were observed. The effects of cytisine on respiration and blood pressure were blocked. Also no direct effect on muscarinic receptors was found; larger doses have spasmolytic nonspecific effects on intestinal smooth muscle. Small doses caused contraction. The alkaloids depressed the motor activity of experimental animals, but in a large dose (150 mg/kg) they caused a tremor. The hypnotic effects of barbiturates were potentiated. When given in the vertebral artery their effects were more pronounced than when given i. v. They blocked the

pressor reflexes evoked by pressing the carotid artery or by

stimulation of the proximal part of cut sciatic nerve. They

blocked the polysynaptic facilitation of the knee-phenomenon

after stimulation of the central end of a cut contralateral scia-

tic nerve, but not when the ipsilateral sciatic nerve was stimu-

lated (monosynaptic impediment). They inhibited the respira-

tion of the rat cerebral tissue and the activity of the rat brain

cholinesterase. The subcutaneous LD_{50} in the mouse was 237. 9

mg/kg.

VI. PHARMACOLOGY OF VINCA ERECTA

Many alkaloids have been isolated and tested by a group of

authors from Soviet Uzbekistan. The most frequently describ-

ed alkaloid with a very different pharmacological action is

vincanine (synonymous with norfluorocurarine, found also in V.

herbacea) with strychnine-like properties. Sultanov in 1959[52]

estimated the lethal dose (LD_{100}) of the chloride to be 7 mg/kg

in rabbits, which is about 10-12 times more than the LD_{100} of

strychnine in the same species. The character of intoxication

and of seizures (tetanic, evoked by external stimuli) was the

same as for strychnine. The effects were antagonized by chloral hydrate. In 1961[53] he, with Egorova, studied the effects of vincanine on cholinesterase in the blood serum of rabbits in comparison with the effects of strychnine. In toxic doses, both drugs increased cholinesterase activity. In 1962[54] he continued the study comparing vincanine, its quaternary base, and vincanidine on rabbit jejunum. In contrast to the stimulating effect of strychnine, vincanine showed papaverine-like properties, while vincanidine and vincanine derivatives exhibited an adrenolytic effect in concentrations of 1:10, 000 - 1:2, 500.

Structure-activity relationships were followed in derivatives of vincanine by Vakhabov and Sultanov[55, 56, 57]. Feline sympathetic ganglia and neuromuscular junctions were used. Methyl, ethyl, butyl, propyl, and amyl derivatives were compared. The subcutaneous LD_{50} in mice increased with an increasing length of the side chain (methyl-79 mg/kg, ethyl-130 mg/kg, propyl-168 kg, butyl-230 mg/kg, amyl-184 mg/kg). In the sympathetic ganglia the methyl derivative was the most effective, while on the neuromuscular junction the amyl derivative (cats i. v.) was most effective. An encephalographic investigation[58] in rabbits after s. c. administration of 2-6 mg/kg showed changes similar to those seen with strychnine in doses of 0. 25-0. 5 mg/kg.

Favorable clinical trials were performed in patients with

neurological indications with vincanine hydrochloride[59] as an

analeptic.

Saidkasmov[60, 61] studied the pharmacological properties

of ervinine (synonymous with kopsinine). The alkaloid poten-

tiated the effects of corazol, strychnine, camphor, caffeine,

and vincamine in mice. Its own analeptic effect was evoked

by subcutaneous administration of doses between 100-200 mg/kg.

In later work[62] he compared kopsinine with pseudokopsi-

nine in rabbits, mice, dogs and cats. The subcutaneous LD_{50}

in mice was 125 mg/kg, i. v., and the LD_{100} in rabbits was 20

mg/kg. Kopsinine at a dose of 1-2 mg/kg i. v. stimulated the

respiration and increased the blood pressure for 20-30 minu-

tes. Higher doses caused a fall of blood pressure and paraly-

sis of respiration. Pseudokopsinine was about twice as toxic.

Its effects on respiration and blood pressure were, however,

less pronounced.

Tombozine (synonymous with normacusine-B), another al-

kaloid from V. erecta, showed a sedative effect[63] at doses of

20-50 mg/kg in mice, and in cats and dogs 0. 05-5. 0 mg/kg

caused a drop in blood pressure. This effect was not blocked

by atropine (1 mg/kg). In higher doses (20 mg/kg) a ganglionic

blockade was observed, as was a depression of the heart. The

LD_{50} in mice (s.c.) was 325 mg/kg and (i.v.) 65 mg/kg. The

papaverine-like effects on rabbit jejunum at concentrations of

1×10^{-4} to 1×10^{-5} were similar to vincanine.

Vineridine[64], a slightly less toxic alkaloid (LD_{50} in mice

i.p. 485 mg/kg, i.v. 125 mg/kg), caused sedation at a dose of

10 mg/kg in mice and relaxed the smooth muscle of the rabbit

jejunum when diluted $1:10^{5}$. Rabbit uterus and cat uterus <u>in</u>

<u>vivo</u> responded to the same dose with contraction. Doses of

1-10 mg/kg caused stimulation of respiration in cats and a fall

of blood pressure. The sympathetic ganglia were partly blocked.

Vinervine[65] was effective in lowering the blood pressure of

rabbits with hypertension, which persisted for four to seven

hours after 0.1-2.0 mg/kg i.v. No effects were observed in

normal rabbits. In cats and dogs, doses of 1-10 mg/kg were ef-

fective as a hypotensive in two phases, similar to the two pha-

ses of hypotension with vincamine (short one, followed by a pro-

longed one). The LD_{50} in mice was i.v. 24.5 mg/kg, i.p. 100

mg/kg.

Vincarine[66], another alkaloid with sedative effects in mice,

rats and rabbits, was effective at doses of 20-40 mg/kg. The

LD_{50} in mice was i.p. 330 mg/kg s.c. 520 mg/kg.

Ervamidine[67] (synonymous with akuammidine) showed se-

dative, central muscle relaxing and hypotensive properties at doses of 0.5 mg - 5 mg/kg in rabbits, cats, and dogs. At doses

of 5-50 mg/kg the alkaloid was ganglioblocking and centrally

adrenolytic. Its effects on the EEG were compared with reser-

pine, vincamine and ervamine. The intraperitoneal LD_{50} in

mice was 391 mg/kg, s.c. 550 mg/kg.

Another alkaloid studied was ervamine[68-70], which has an

LD_{50} in mice (i.v.) of 90 mg/kg, (i.p.) 200 mg/kg and (p.o.)

370 mg/kg. Ervamine evoked sedation in rabbits and dogs,

prolonged the effect of hypnotics, had a short-acting hypoten-

sive effect, stimulated the heart and increased the coronary

flow. It has a pronounced oxytocic effect, as well as a hemo-

static effect. The alkaloid was compared with the total alka-

loids of Vinca erecta and the LD_{50} in rabbits i.v. was 12.9 mg/

kg for the total alkaloids, and 95 mg/kg for ervamine. In toxic

doses both caused clonic seizures. Lethal doses of both cau-

sed similar histological changes: acute hypermia of the brain,

solar plexus, kidneys, myocardium, lungs, spleen, stomach,

intestine, and uterus; there was granular dystrophy of the epi-

thelium of the kidneys and vacuolar degeneration of the nerve

cells of the brain and solar plexus. In rabbit experiments the

subcutaneous injection of the total alkaloids for 20 days, or the

intravenous injection of ervamine at 10 mg/kg/day for 9-17 days

did not exert any toxic action and did not change the morphological structure of the internal organs. In these experiments ervamine caused a 10% reduction in the amount of hemoglobin in the blood while the total alkaloids increased hemoglobin by 13-23%, but caused leukopenia and erythrocytosis.

Other alkaloids with described pharmacological effects are vinerine[71] and vinervinine[72].

VII. <u>SUMMARY</u>

Unfortunately the only group studying the chemistry and pharmacology of <u>V</u>. <u>erecta</u> alkaloids is the Turkestan group who have often published incomplete and difficult to compare data, in journals not readily accessible. On the other hand, the <u>V</u>. <u>erecta</u> alkaloids have been pharmacologically characterized to a greater extent than those of any other <u>Vinca</u> species. <u>Vinca herbacea</u> has been studied pharmacologically only by the Bulgarian group, and the study has not progressed beyond the first stage of an evaluation of total alkaloids. The few French and American papers on <u>V</u>. <u>major</u> also do not go much beyond this point. Thus, only the main alkaloid of <u>V</u>. <u>minor</u>, vincamine, has been well studied pharmacologically, thanks to the competitive experiments of Hungarian, Czechoslovakian, Polish, French and to some extent, Russian authors.

Evidently not only the variation in experimental animals,
but variation in plant material and isolation procedures have
been responsible for differences in results. The clinical results from alkaloids other than vincamine are not beyond the
initial clinical observation stage, and even the studies with
vincamine leave much to be desired. The pharmacological
evaluation of alkaloids from Vinca species still more a task
for the future than a subject of present discussion.

REFERENCES

1. M. Pichon, Bull. Mus. Hist. Nat. 23, 439 (1951).

2. A. Orechoff, H. Gurewitch, S. Norkina and N. Preis,
    Arch. Pharm. 272, 70 (1934).

3. A. Quevauviller, J. LeMen and M. M. Janot, Ann. Pharm.
    Franc. 12, 799 (1954).

4. A. Quevauviller, J. LeMen and M. M. Janot, Compt. Rend.
    Soc. Biol. 148, 1791 (1954).

5. Gazet duChatelier and E. Strasky, Ann. Pharm. Franc.
    14, 677 (1956).

6. J. Hano and J. Maj, Polski Tygodnik Lakarski 11, 3 (1956).

7. J. Hano and J. Maj, Acta Polon. Pharm. 15, 71 (1957).

8. J. Maj, Far. Polska 12, 85 (1956).

9. S. L. Voskoboinik, Farm. Zh. (Kiev) 16, 47 (1961).

10. D. K. Zheliazkov, Survem. Med. (Sofia) 9, 16 (1958).

11. E. Szczeklik, J. Hano, B. Bordanikova and J. Maj, Pol-
    ski Tygodnik Lekarski 12, 121 (1957).

12. Yu. A. Shelekhov, Ref. Zh. Otd. Vypusk Farmakol. Toksikol. 54, 19 (1964).

13. Yu. A. Shelekhov, Tr. Alma-Atinskogo Med. Inst. 21, 320 (1964).

14. Raymond-Hamet, Compt. Rend. Soc. Biol. 148, 1082 (1954).

15. L. Szporny and K. Szász, Arch. Exp. Pathol. Pharmakol. 236, 296 (1959).

16. L. Szporny and P. Gorog, Arch, Int. Pharmacodyn. Ther. 138, 451 (1962).

17. L. Szporny and K. Szász, Acta Physiol. Acad. Sci. Hung. 14, 46 (1958).

18. I. Krejci, Cesk. Fysiol. 8, 452 (1959).

19. P. Gomori, E. Glaz, Z. Szabo, Orvosi Hetilap 101, 361 (1960).

20. Z. Szabo and Z. Nagy, Arzneimittel-Forsch. 10, 811 (1960).

21. I. Mueller, Orvosi Hetilap 102, 359 (1961).

22. A. Kaldor and Z. Szabo, Experientia 16, 547 (1960).

23. J. Machova and L. Macho, Biologia (Bratislava) 17, 456 (1962).

24. P. Gorog and L. Szporny, Biochem. Pharmacol. 11, 165 (1961).

25. P. Gorog and L. Szporny, Biochem. Pharmacol. 8, 259 (1961).

26. O. Linet, I. Krejci and M. Hava, Acta Biol. Med Ger. 9, 158 (1962).

27. J. Molnar and L. Szporny, Acta Physiol. Acad. Sci. Hung. 21, 169 (1962).

28. J. Machova and J. Mokry, Arch. Int. Pharmacodyn. 150, 516 (1964).

29. J. Machova, Arch. Int. Pharmacodyn. Ther. 165, (2), 459 (1967).

30. J. Machova and F. V. Selecky, Bratislav. Lekarske Listy, 43, 449 (1963).

31. A. G. Kurmukov, Farmakol. I. Toksikol 30, 286 (1967).

32. A. G. Kurmukov, Farmakol. Alkaloidov Glikozidov 87-93 (1967).

33. A. G. Kurmukov, Farmakol. Alkaloidov Glikozidov 74-79 (1967).

34. A. G. Kurmukov and M. B. Sultanov, Farmakol. Alkaloidov 2, 171 (1965).

35. F. Kaczmarek, J. Lutomski and T. Wrocincki, Biul. Inst. Roslin Leczniczych 8, 12 (1962).

36. A. G. B. Kovach, P. Sandor, S. Biro, E. Koltay, K. Mazan, O. Caluder, Acta Physiol. Acad. Sci. Hung. 36, 307 (1969).

37. Z. Szabo, Z. Nagy and E. Kiss, Magy. Belorv. Arch. 15, 186 (1962).

38. E. Ezer and L. Szporny, Kiserletter Orcostudomany, 19, 67 (1967).

39. L. Szporny and P. Gorog, Conf. Hung. Therap. Invest. Pharmacol. 2, Budapest pp. 237-240 (1964).

40. F. Hargitai and J. Werkner, Orvosi Hetilap 103, 312 (1962).

41. F. Solti, M. Iskum, A. Peter, J. Rev. R. Hermann and K. Foldessy, Cor et Vasa 6, (2), 138 (1964).

42. F. Solti, Orvosi Hetilap 103, 202 (1962).

43. A. Szabor and G. Imre, Paper read at the Third Hungarian Conference of Therapy and Pharmacol. Res. October 6-11. (1964).

44. M. Foldi, F. Obal and G. Szeghy, Med. Welt 37, 2122 (1965).

45. A. Ravina, Presse Med. 74, 525 (1966).

46. A. Quevauviller, J. LeMen and M.M. Janot, Ann, Pharm. Franc. 13, 328 (1955).

47. A. Quevauviller, M.O. Blampin, Pathol. Biol. 6, 1481 (1958).

48. N.R. Farnsworth, F.J. Draus, R.W. Sager and J.A. Bianculli, J. Am. Pharm. Assoc. Sci. Ed. 49, 589 (1960).

49. K.S. Roussinov, D.J. Zhelyazkov and V.P. Georgiev, Arch. Ital. Sci. Farmacol. 11, 83 (1961).

50. K. Roussinov, D. Zhelyazkov and V. Georgiev, Compt. Rend. Acad. Bulg. Sci. 15, 329 (1962).

51. K. Roussinov, D. Zhelyazkov and V. Georgiev, B'lgarska Akad. Nauk. Otdel. za Med., Nauk 5, 27 (1962).

52. M.B. Sultanov, Izvest. Akad. Nauk, Uz. SSR, Ser. Med. 3, 38 (1959).

53. M.B. Sultanov and T.A. Egorova, Dokl. Akad. Nauk. Uz. SSR. 10, 28 (1961).

54. M.B. Sultanov and E.B. Baibekov, Ref. Zh. Otdel. Vypusk Farmakol. Toksikol. 54, 121 (1962).

55. A.A. Vakhabov and M.B. Sultanov, Akad. Nauk Uz. SSR, Khim.-Tekh. i. Biol. Otd. pp. 30-33 (1966).

56. A.A. Vakhabov and M.B. Sultanov, Farmakol. Alkaloidov, 2, 183 (1965).

57. A.A. Vakhabov and M.B. Sultanov, Akad Nauk. Uz. SSR, Tashkent 1, 133 (1962).

58. A.G. Kurmukov and M.B. Sultanov, Akad. Nauk Uz. SSR Uz Biol. Zh. 1, 32 (1967).

59. Sh. Shamansurov and M.B. Sultanov, Zh. Nevropatol. Psikhiat. Im. S.S. Korsakova 67, 1807 (1967).

60. T. Saidakasymov and M. B. Sultanov, Akad. Nauk Uz. SSR. Otd. Geol. -Khim. Nauk pp. 176-180 (1961).

61. T. Saidkasymov, Voprosy Biol. i Kraevoi Med. Akad. Nauk Uz. SSR, Otdel. Biol. Nauk pp. 242-247 (1960).

62. T. Saidkasymov, M. B. Sultanov and T. A. Egorova, Akad. Nauk Uz. SSR, Khim. Biol. Otd. pp. 33-36 (1966).

63. M. B. Sultanov, T. Saidkasymov, Dokl. Akad. Nauk Uz. SSR 22, 41 (1965).

64. A. K. Kurmukov and M. B. Sultanov, Akad. Nauk Uz. SSR Khim-Tekh. i Biol. Otd. pp. 26-30 (1966).

65. A. K. Kurmukov, Med. Zhur. Uz. 6, 49 (1967).

66. M. Khanov, A. K. Kurmukov, M. B. Sultanov and Kh. S. Akhmedkhodzhaeva, Farmakol. Toksikol. 31, 47 (1968).

67. A. G. Kurmukov, Farmakol. Toksikol. 31, 47 (1968).

68. M. B. Sultanov and T. Saidkasymov, Farmakol. Alk. (Tashkent) 2, 128 (1965).

69. A. G. Kurmukov, L. N. Goldberg, M. B. Sultanov, L. A. Vysotskaya and A. A. Kulikov, Vop. Morfol. Nekot. Zabol. Uzb. pp. 157-161 (1965).

70. A. G. Kurmukov, L. N. Goldberg, M. B. Sultanov, L. A. Vysotskaya and A. A. Kulikov, Ref. Zh. Otd. Vyp. Farmakol. Khimioter., Sredstva, Toksikol. 3, 490 (1967).

71. A. G. Kurmukov, Farmakol. Alkaloidov Glikozidov, pp. 83-87 (1967).

72. A. G. Kurmukov and Kh. S. Akhmedkhodzhaeva, Farmakol. Alkaloidov Glikozidov pp. 79-82 (1967).

AUTHOR INDEX

Numbers in brackets are reference numbers and indicate that an
author's work is referred to although his name is not cited in the
text. Underlined numbers give the page on which the complete
reference is listed.

A

Abbott, B. J. , (19) 148 , (209) 160
Abdurakhimova, N. , (63)243,
    (229, 230) 161, (331, 332) 168,
    (348, 358) 170, (12, 20) 277
Abdychakumov, A. A.,(319) 168
Abidov, A A., (319) 168
Abraham, D. J.,(294)166
Achelis, J. D. , (258, 259) 163
Aghoramurthy, K. , (225) 161
Ahmed, Q. A. , (135), 155
Akhmedkhodzhaeva, Kh. S. ,
    (66, 72) 338
Aldaba, L. , (8) 147
Aliev, A. M. , (355, 356) 170
Adityachaudhury, N. (133)155
Amai, R. L. S. (44) 242
Ang, S. K. (46) 242
Antonaccio, L. D. (131) 155 ,
    (26) 240
Antonaccio, L. N. , (128) 155
Aripov, Kh. N. , (300) 166 ,
    (322) 169 , (10) 277
Arnaud, M. , (215) 160
Arndt, R. R. , (255) 163
Arthur, H. R (275) 164
Atal, C. K. , (158) 156, (247) 162

B

Babaev, N. A. , (354-356) 170
Bailiekov, E. B., (306) 167
Banes, D., (152) 156
Bartlett, M. F. , (115, 116) 154,
    (18-20) 240, (44, 48) 242
Barton, J. E. D. , (23)240, (44)242
Basu, N. K. , (188) 159

Bate-Smith, E C. , (214) 160
Batkiewicz, E. , (237) 162
Batllori, L. , (185) 158
Battersby, A. R. , (47)242,(7)277
Bauer, S. , (90) 152
Bauerova, O., (90) 152
Bayer, J. , (110)154 , (2) 302
Beak, P. , (44) 242
Beale, G. H. , (57) 150
Beereboom, J. J. , (160) 156
Bein, H. J. , (170)156, (212) 160
Belikov, A. S. , (118) 154
Bernal, R. M., (144) 156
Beugelmans, R., (232) 162 ,
    (52) 242
Bevan, C. W, L. , (68) 244
Bhatnagar, J. N. , (190) 159
Bhattacharji, S. , (248) 163
Bianculli, J. A. , (75) 151
Biemann, K. , (114, 122) 154,
    (143) 156 , (195) 159 , (273) 164,
    (25) 240, (32, 40) 241, (49, 51)
    242
Biro, S. , (36) 336
Birosyak, Yu. G. , (79, 80) 152
Bisset, N. G. , (2) 147
Bittner, E. , (78) 151
Blaha, K. , (70, 71) 151, (14) 240 ,
    (35) 241, (89) 152
Blazek, Z. , (43-45) 149
Blomster, R N. , (210) 160 ,
    (281) 165 , (294) 166 , (295)166
Blossey, E. C., (253) 163
Bocharova, D. A. , (282) 165,
    (320) 168
Bodendorf, K. , (269) 164
Boderick, D. B. , (51) 150
Boegemann, W.H. , (179) 158

339

60177